Zentanglements - The Three Laws Of Consciousness For Smarties

Michael Mathiesen

Published by Michael Mathiesen, 2021.

Preface

"No problem can be solved from the same level of consciousness that created it." Albert Einstein

---

THIS BOOK IS DEDICATED to using the wisdom of Professor Einstein to save this planet using a new way of looking at the Cosmos through Zentanglements and the Three Laws of Consciousness.

First of all, I feel the need to define precisely what a Zentanglement is. It's a new word that I have invented and it combines the word 'Zen' with 'Entanglement'. Zen being a frame of mind for meditation that goes back many centuries and began in Buddhist India. It speaks about finding the most basic and fundamental truth about any given situation.

When I thought about this concept the other day, I realized that Zen is almost exactly like Science because both of these endeavors are out to find the most basic and fundamental truth about the universe. At this time, I also became aware of a new form of art called 'Zentangles' and it's a nice and simple method for anyone to turn a doodle into an intriguing and thoughtful work of art while also relieving the stress and anxiety of modern living.

So I bought myself a sketch pad and some art pencils, poster board and pastel chalk crayons and gave it a try. Before long I had some simple designs that illustrated my long-held philosophies of art, music and Science, sometimes even better than words.

For a long-time writer this astounded me. You will see many of my initial Zentangled doodles throughout this book, which turns out to be the culmination of my life-long work in Quantum Physics and Cosmology.

The old adage seems to hold true in that a 'Picture is worth a thousand words'. If that's true, then what you have before you is a very scaled-down and economical treatise with few words, but also a much deeper exposition of my ideas in these Zentanglements that we may all come to know and love.

And, I've put in the title that the book is intended for 'Smarties', but only because 'For Dummies' didn't seem appropriate when considering some of the most vexing problems of our times. And it also seems to me that we have too many 'Dummies' in charge of the world right now and so this book is also an appeal to everyone to up their game a notch or two if we're serious about surviving this century.

But, I promise you do not have to be that smart to completely understand these ideas. Einstein's greatest contribution to the world is not his genius in unmasking so much of the universe, but that he was able to do so in ways that most of us can and do understand.

Thus, I give you 'Zentanglements'.

And since these words and images are designed to go together, you can also find them for downloading or ordering prints and canvases at my website – Zentanglements.com

Introduction

I Get A 'Big Bang' Out Of You

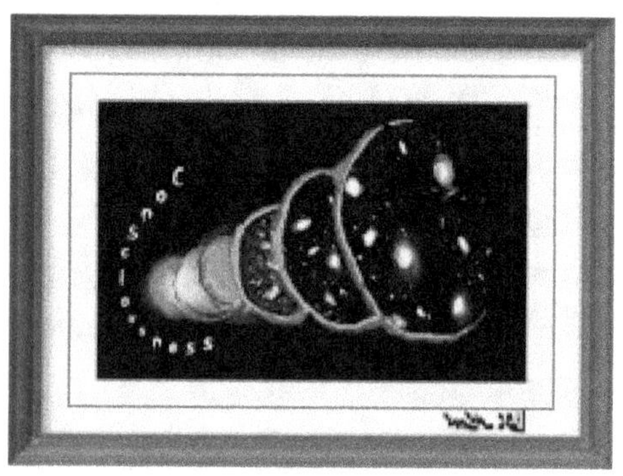

———◦———

ALTHOUGH YOU ARE ABOUT to be ploughing through some pretty big concepts, the Science I present here is easy enough for anyone with a High School education to understand and follow. All you need is the desire to know more about these things. If you are only interested in keeping up with what the celebrities have been doing, these concepts won't adhere to your brain, which by the way, can become polluted and rendered a toxic waste dump by the over-indulgence in the nonsense presented to us all in our modern world. This is the 'dumbing down' of America that we all can see happening with our own eyes, if we are not yet in the totally dumbed down crowd.

Therefore, the necessity for this book that I have geared for the 'Smarties' in the world, not the 'Dummies'. So, if you are a 'Dummy', you actually need to read this book more than anyone else, because as long as you have breath, it is not too late for your brain to wake

4

up. If you're a 'Smartie', on the other hand, you're way ahead of the 'Dummies' already and you will want to read this book to stay ahead. We need every one of you to bring your particular point of view to the party, as you will see.

———◉———

AS YOU MAY RECALL FROM your readings and discussions over the years, Einstein's Theory of Relativity has been proven many times over. It basically says that everything in the universe is relative to its speed. For example, we all know that light is the fastest thing that we know about in the universe and we even measure the size of the universe by how long it takes a beam of light to travel at its maximum speed of 300,000 Km/Sec.

We can also easily remember the most famous equation in history – Einstein's E=mc2, which states that all energy (E) and matter (M) are really the same because they can be transformed one to the other multiplied by rate of the speed of light (c) squared, a huge number, almost unimaginable for our tiny brains to comprehend – 300,000 km/second X 300,000 km/second or around 90 billion kilometers per second. Now, that's bookin' it.

But, the really incredible part of Einstein's theory that most of us rarely consider is that Time stops at the speed of light. This means that if you are a photon of light traveling from the nearest stars, around 5 light years away, toward the earth, it means for 5 years Time has stopped for that particular little particle of light because it is traveling at the speed of light and Einstein's theory of relativity makes it impossible for anything else to happen.

So, as I was having my inspiration for this book, I imagined what it would be like to be a photon of light moving along in Space and Time but from the photon's perspective where all of Time has stopped.

If a photon could talk, what would it reveal to us? His universe shows no movement. He is the only object moving. Since all of Time has stopped for this little critter, nothing else around him seems to be moving at all. In effect, he's moving through a still-photograph of the universe. His only mission is to arrive somewhere else in the still photograph which remains the same as the moment he was born inside a star.

At that moment many years in his future, he will land on your retina or my retina or anyone else's retina who is looking up at the night sky in his direction.

But, it gets better. The very next instant that we use our eyes to look up at the night sky, another photon arrives upon our retina and gives us the snapshot of the universe, one millionth of a millisecond later. Then, another photon lands on our retina and the same thing happens, perhaps thousands of times, before we can even blink. This results in our perception that the universe is a series of still photographs taken by each photon of light at the time of their departure, but the arrival of the photons are so fast and furious upon our retinas that it creates the illusion of a moving picture, a movie that represents the background of all our lives.

I take this to mean that the Creator invented a single particle in Nature that serves as the messenger of information to the rest of the universe. Since all of Time stands still from the relative position of the light beam, it's a pretty simple thing for the photons of light to carry the information of what everything looks like in a very wide-angle lense at any given instant of time. Luckily there are more than one photon speeding around the universe giving us this constant reporting. There are literally an infinite number of photons hitting our eyes every second, so as to produce the illusion of Time moving forward minute by minute, hour by hour, day by day, year after year, counting down on the gradually diminishing amount of time we have left to enjoy our impacts here in this Space.

In reality, Time does not exist, or it exists in short bursts that flicker along so fast that it convinces us that something is alive and moving. It's only living things, if the truth were known, that exists in this universe with any sense of Time, because our lives start heading for our exit from this place from the very moment that we arrive here and there is a little hour-glass inside our brains that keeps track. 'Life is short', as they say.

At the time of our birth, we come out of the great expanse of Time that has no starting point, ending point or movement of any kind according to the messengers of Time, the photons of light that originate inside of the Stars.

When we pass from this place, we return to the infinite existence of Time without movement. And as soon as we are ONE with the universe we probably become completely aware of the true nature of the universe where Time creates everything by reporting the changes through the method of light transmission, the tiniest messengers sending us constant updates of all the changes in the Consciousness of God.

In other words, Time is not what's going on in the real world. The universe is merely a thought process from something that processes thoughts much more powerfully than we can and this is what we like to call - Cosmic Consciousness or the Consciousness of God.

———◉———

BEFORE I END THIS SECTION, you should also know that the word - 'Zentanglements' is also referring to the concept of 'Entanglements' in Quantum Physics, a new discovery that showed us that electrons, photons and probably all other subatomic particles can become 'Entangled' with one another in a pairing, not unlike the way all living things pair up in order to produce the next generation, the new and improved human.

The Entangled electrons, or photons start to act in opposite relation to one another. They spin in the opposite direction from one another and this is where I am always reminded of the symbol for Yin and Yang.

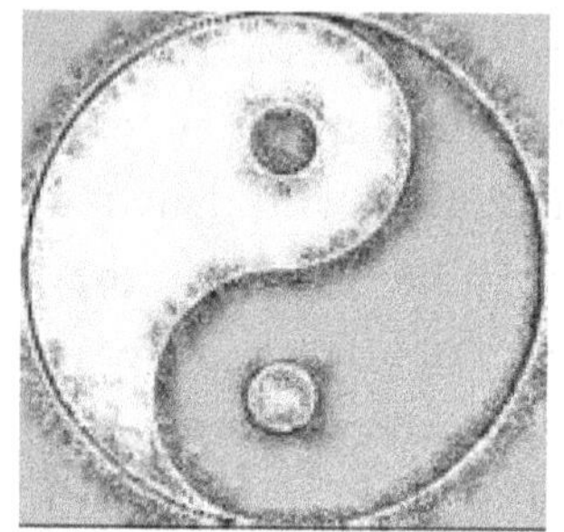

My version of Entangled Electrons

During the time of the creation of this age-old symbol, no one had ever heard of Entangled Electrons or even anything much about Atoms either. Yet, someone came up with this little doodle that we all know and it's amazing because it looks as if it were drawn by a modern-day Physicist who is trying to explain how electrons become entangled and how one spins in an UPWARDS DIRECTION and the other entangled electron spins DOWN.

My Yin-Yang Zentangled

BUT, THE TRULY AMAZING mystery about Entangled electrons or photons is that when we separate the two particles, even by thousands of miles and we alter the spin of one of them, the other particle that is entangled immediately spins in the opposite direction in order to remain in a tight bond of some kind with its partner. Bear in mind, there is no passage of time, not even a microsecond, therefore, there is some other hidden force that supersedes all of Space and Time and even the venerable theory of Einstein's that nothing can travel faster than the speed of light and everything else is relative to that event.

Somehow, both Entangled Electrons or particles are aware of the other one's state and are able to reverse their own state of existence to maintain the condition that resembles 'Yin and Yang' where one spins UP and the other spins DOWN.

When I heard about this phenomenon of subatomic particles who mate for life like swans and faithfully act together in perfect synchronization of their lives once they bonded together, it made me realize something very important about Consciousness. In fact, this was and still is the greatest evidence ever produced by anyone that Consciousness exists all over the place and it's not just entrusted inside of the human brain and nowhere else.

Having owned several dogs and cats in my life, I already had a feeling that humans were not the only participants in Consciousness. And, probably most of us feel that most animals have this same feature in life. But, this bit of Science that shows how electrons and photons and probably all other subatomic particles can be aware of each other showed me something far more amazing than just dogs and cats could show us.

And, when you look at other aspects of subatomic physics it seems highly likely that just about every form of physical existence in the universe including insects, bacteria, viruses, trees and even quarks, protons and leptons all share a common medium of Consciousness. After all, every single electron, every single proton, every single neutron, baryon, quark, or lepton, of which there are countless trillions in every cubic millimeter, are all exactly the same. They all have the same size and shape and all share the same exact amount of energy. The human mind is not capable of imagining how many electrons or other subatomic particles that exist in the universe. However, we do know that these countless numbers of them are all the same. Out of what kind of cookie cutter do they originate? Where is the amazing 'zero-defects in manufacturing' factory of subatomic particles?

Before going any further, let's all recall how I started this segment with a reminder about Einstein's theory of Relativity. And so now it's time to ask you that If Time can stop whenever you reach the speed of light - what is Time, the only thing in the universe that can stop completely in its tracks and without any head snaps?

This book is a humble attempt to bring us all closer to the light.

One of my good friends just happens to be a world-renowned Astro-physicist and several years ago, she invited me to one of her lecture series at the local University.

Towards the end of her lecture series on the Evolution of the Universe, she had a picture on the wall of the first few seconds after The Big Bang. Of course, she described the Big Bang as the start of the universe. According to all current Scientific knowledge there was nothing prior to the Big Bang. Then, suddenly, there was this largest explosion in history and everything is created in that explosion, at least all of the electrons, protons, photons, quarks, etc. that will be required to clump together to eventually become the stars, the planets, the galaxies that we see in the night sky. All of it is created at the moment of the Big Bang.

And so, the Scientific Story of the beginning of everything is not unlike how the Bible describes the beginning of Time.

The opening scene of the Bible in the Book of Genesis:

"Now the earth was formless and void, and darkness was over the surface of the deep. And the Spirit of God was hovering over the surface of the waters. And God said, "Let there be light and there was light. And God saw the light was good, and He separated the light from the darkness.""

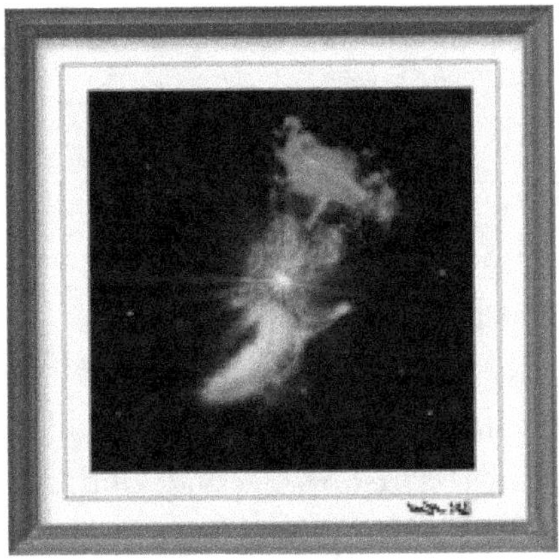

Of course the way that the authors of the Bible put it is circular, because they say that the Earth was already formed, just dark, like a theater is dark before the lights come on and the show begins. In reality in the opening of the universe, there was no Earth in any kind of shape or form. In fact, the Earth would be formed some 10 billions years or so after all of this light happened during the Big Bang.

Toward the end of her lecture series, my friend the famous Astrophysicist has a diagram of the first few seconds of the Big Bang. It looks like this. It is meant to illustrate the first few seconds of the origin of the universe, and it's all gathered from real data that astronomers have discovered about our universe over the centuries of staring at the light glowing from the most distant stars and even from the sound of the Big Bang that still exists at about three degrees Kelvin everywhere we look and one of the biggest pieces of evidence that the Big Bang did in fact happen.

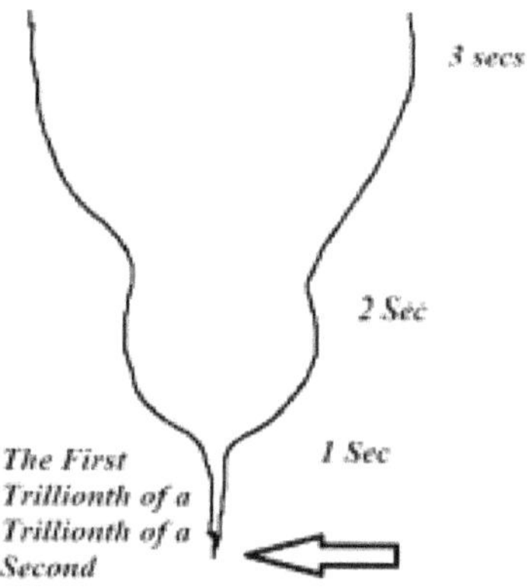

My friend goes on to describe the Diagram as representing the first few seconds of the Big Bang and how it is expanding out to become the entire universe very fast in what is known as the Great Expansion. THEN, SHE POINTS TO the place at the very bottom point of this Time Chart of the Big Bang and she says:

"At the first trillionth of a trillionth of a second of the Big Bang, all the information the universe needed was imparted."

I raised my hand and asked her. "Jean, did you really mean to use the word 'Information'?

She nodded her head very emphatically and said 'Yes', and then moved on with her lecture.

For me, this was HEADLINE NEWS OF THE CENTURY, or possibly of all centuries. I couldn't believe it. I looked around at the other people, mainly Astronomy students, sitting in the classroom and I tried to pick out how many of them may have truly understood the significance of this answer.

I noticed that only one or two percent were whispering among themselves in some kind of recognition. But for me, this was one of the most unforgettable moments in my life because my friend, who I had always regarded as an Atheist had just laid down how God created the universe, just as it is implied in the Bible, because who other than God would hold all the information that the universe would need over the next 13.7 billion years inside the space of a trillionth of a trillionth of a second and deliver it this way so that it could all be spread over the rest of the universe so uniformly?

I believe you have to pause here for a moment to consider and truly appreciate this amazing truth being taught in Astrophysics today before you can absorb the rest of this book. Later, my friend would tell me that by 'Information' she meant all of the Laws of Physics were imparted in that first trillionth of a trillionth of a second, but either way, whatever you want to call it, the laws of Physics must be based on trillions upon trillions of pieces of data, at least in the human world I know about. The Laws of Physics have to be created out of some Consciousness that the universe must be composed so that the Laws can come out of that experience.

This Scientific truth also implied to me that all of the information that universe is based on must have existed in another space of Time

PRIOR to the Big Bang, because how could all of that information be created in the space of a trillionth of a trillionth of a second. We're talking about all of the information about how to create electrons, protons, quarks etc, how they can form into atoms and atoms into molecules, all of the laws relating to molecular energy.

Remember any Science Course or Math course in your life, how the books they gave you at school were packed with scientific facts and numbers. Remember Avogardro's Number? All of that information plus all the other courses you've taken and not taken in your life, plus all of the information held in the billions of books kept in all the libraries of our planet over thousands of years.

Add to this huge pile of information, all of the information that we don't even know about yet. Add to this all of the information stored in books and computers on other planets, of which there must be billions of them. The amount of information this universe requires is staggering, to say the least, and couldn't possibly be created in a trillionth of a trillionth of a second. That seems absurd, right?

Unless some kind of storehouse of all that information exists somewhere outside of this universe, I'm thinking. This storehouse of all this information could be a huge computer chip somewhere or more likely it's a form of a higher state of Consciousness where feats like this are mediocre and humdrum. If it's a huge computer chip or other information storage device, it begs the question of who manufactured that computer chip or other storage device? It's certainly not Intel or any other Earthly chip maker, got that?

My Astrophysicist friend went on to say that the reason we know this is because the universe expanded out at a speed greater than the speed of light. And, she pointed out that all of the stars and galaxies at one end of the universe behave exactly like the stars and galaxies at the opposite end of the universe. Since nothing can travel faster than the speed of light, then, the two opposing sides of the universe could never have communicated with each other and therefore all of the information required to create the universe had to have been produced at the very earliest moments in the Big Bang. That way, all parts of the universe would contain all the same information.

That made sense to me at the time. But, I wasn't thinking, nor was she going to comment about Einstein's theory of relativity in this regard - because how is the universe created as it is expanding at the speed of light and all of Time stops at the speed of light? And how could the universe itself expand out faster than the speed of light when 'Relativity' says that nothing can travel faster than light?

Did the universe acquire a special exemption permit from the office of Universal Expansion to go faster than what is legally allowed everywhere else and thus defying the Laws of Physics that it had created in the very beginning? As Winston Churchill once commented: "It's a riddle wrapped in a mystery inside an enigma."

Just for fun, let's say that the evidence regarding the Big Bang that we've acquired is all correct and the universe did expand faster than the speed of light at one point, for a while. Since all Time stops when you reach the speed of light, what happened to the age of the universe during the Time that the clock was stopped? Did it advance at all? So, could our universe be much older than we think because of the period of Time when Time stopped? How long do you suppose it stopped? And why?

And, if the universe expanded faster than the speed of light, what happens to Time when you exceed the speed of light? Does Time flow backwards? If so, does Time when it reverses itself cancel out more time than it would acquire by going forward? In other words, can Time not only stop, but accelerate backwards? Could the age of our universe be double, triple or even ten times what we can measure since Time was flowing backwards and for an undetermined amount of it?

I'm just following this line of thinking that has hit my tired old brain since I learned some of the facts about the Big Bang. Don't worry, I'm going to solve these problems for you by the end of this book because in this book, I'm going to tie up all the loose ends of Cosmology for you and you'll be very happy you spent this amount of time with me in this quest and I promise it won't run backwards.

Chapter One

The God Particles

We've covered much ground so far, though you may not know it. In effect, if you carefully read the Introduction, you are now on the cutting edge of Astrophysics and Quantum Physics. You now know more than Einstein did when he was alive, because at the time of the death of the greatest mind in history, 1955, we had not yet discovered the God Particle. There were indications of Entanglements in the universe, but no one really understood what it meant for the rest of the laws of Physics.

Einstein's desk on the day he died.

EINSTEIN DID GO TO his reward knowing that his theory of Relativity had been proven. He also knew that his most famous equation, E=mc2, the fact that energy and matter are really one thing, had been proven by the dropping of two atomic bombs on Japan which ended World War II.

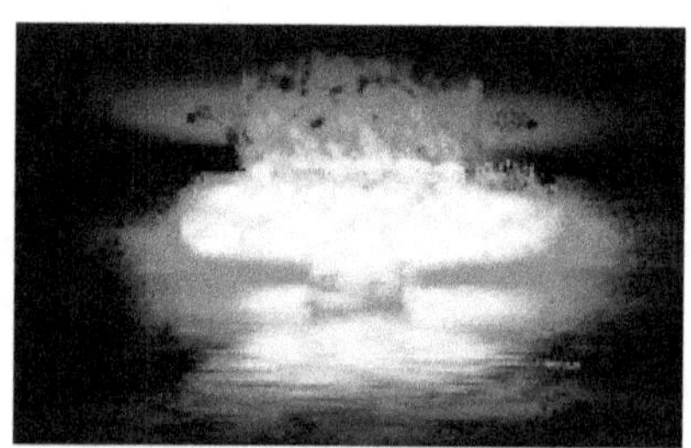

The energy of many thousands of tons of explosive were created by a single pound of Uranium isotope through the invention of the Atomic Bomb, changing life on Earth forever. Now the Hydrogen Bomb, about one thousand times more destructive than the crude Atomic Bombs of World War II, never used in War yet, uses these smaller Atomic Bombs as detonators.

IF WE EVER UNLEASH more than a few Hydrogen bombs on the Earth in the next World War, it will mean the end of everything on the Earth for millions of years.

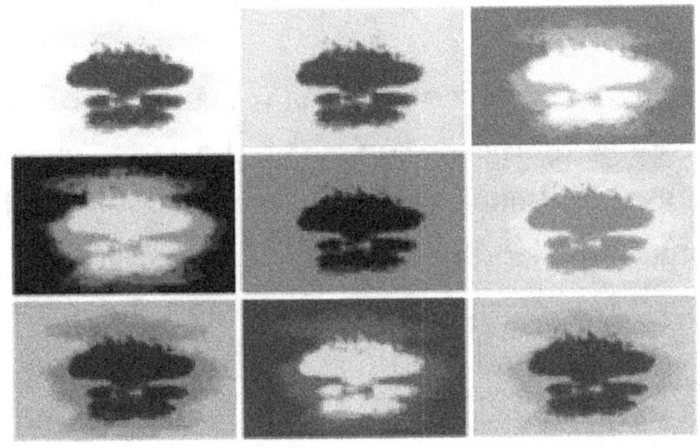

Remember, this will all make sense by the end of the book. Have no fear.
ON THE OTHER HAND, we may have the God Particles to thank for keeping tempers cool, so far, in our government leaders so that they may be kept in storage forever.

How can I make this claim? Because the God Particles are the best evidence so far that God exists, wants us to survive for a little bit longer and may even want to be part of our story.

At the very least, by the end of this chapter, you will hold more important facts about the universe, and the way God works his little miracles, than Einstein even dreamed about when he passed from this place.

Therefore, I want to start out telling you about the God Particle.

It all started with the discovery of the God Particle and how everything was created out of the Big Bang with God using all of his littlest messengers, the God Particles to create everything else.

Then, in rapid succession, we learned more about the concept of 'Entangled Electrons' and Photons. We learned how these most fundamental building blocks of life actually partner up and then get to know about the other partner, in the subatomic realm, and never give up on their loyalty to one another. As soon as I learned about these two recent discoveries of the basis of Nature, I was off on the greatest intellectual mystery adventure of my life or perhaps any other life of our kind.

I hope you can appreciate the confluence of these two new concepts that leapt into life in the last few years. First there were the God Particles that gave every other particle in the universe their ability to size up, to have mass, as the Physicists say, but in reality, they gain a certain size that is uniform and limited in scope throughout the universe all in the same way.

Then, some of these particles that God made for us, are introduced to each other, have their magic moment of deep attraction and the perfect chemistry and fall in love with one another, never abandoning their partner, always faithful and demonstrating their love for one another by changing their state to mirror the state of their significant other, no matter how near or how far, soul mates to the end of time.

Then, someone told me they were drawing Zentangles to relax and bring themselves into a deeper meditative state in order to reach a better creative experience. This is something that I hope to achieve more and more in my writing and other artistic sides of my life, so I had to give it a try.

You should know first that a Zentangle is usually a very simple drawing that a person draws in probably the most contemplative state that the human mind can attain. It's like a doodle, except you want to be thinking about the symmetry in the universe while drawing the doodle and bringing that energy to your drawing.

So, below is a Zentangle of mine that I drew while thinking about the God Particle.

I PICTURE THE GOD PARTICLE in this fashion because matter doesn't exist as particles, only wave actions, just like the surface of the ocean. And the God Particle is no different. However, the persistence of the wave action at any given point makes these tiniest of things in the universe behave like particles and that's why they get their names.

Also the God Particle is the single greatest force in the universe because it gives all the other energies in the universe, their mass. Without the property of 'mass', nothing can exist as we know it. The universe would be a very warm soup of energies that never coalesce into any of the basic fundamental building blocks of the universe and so without the God Particle, there is no universe, at least no universe as we know it.

It's important to know that when they say that the God Particle gives every other particle its mass, this is a misnomer because there is no such thing as 'mass' in the universe, as no solid particle truly exists at this level. Everything is really made out of waves of energy that have such a persistence over time that at this level of the universe, we are fooled by primitive sense organs, into perceiving the weight or the 'mass' of things. And, so the God Particle does NOT render everything its mass. It merely determines its energy packet size. And, these energy packets are all the same everywhere. With an accuracy down to several hundred decimal points, every particle in the universe of the same class all hold the exact same amount of energy and this is due to their interaction with the God Particles or more accurately, the God Particle Field, because the God Particle itself is also an energy.

And so it follows that all electrons have a certain amount of energy that is exactly like all the other electrons that have ever been created in the universe over the many eons of Time. Protons have a certain energy that is also like all the other protons that have ever been created. Quarks all have the same energy as all of their brethren. Muons all have the same energy as all the other muons. Gluons all have the same energy as all the other Gluons. Etc. Right on down the line for all things within the subatomic world, all of it due to their interactions with the God Particle Field, which is everywhere. Thank God!

We will get into the mathematical proof of this shortly.

For now, You should be asking yourself, 'How did a Zentangle make such a difference in the author's appreciation of the God Particle?'

And this is a difficult question for me to answer because it has to do with the 'Creativity' that lies within the God Particle, the Creative

Force that created the entire universe and the tiniest little speck of creativity that enabled me to 'perceive' the God Particle in my mind and translate that vague notion into a diagram that you see above.

If you have ever performed something that is considered to be a creative act, like played music for someone, written a song, written a book, painted a painting, choreographed a dance, designed a new product, you know the creative force. We all have some creative force within us. Some more than others. The ones that exhibit the greatest creativity we idolize and treat as heroes.

Picasso, for example, and in my opinion the greatest artist of all time, we can all agree had boatloads of the creative force inside him, much more supposedly than you or I.

Count them: 4 strokes of the pen - total genius.

EVEN THOUGH HE WOULD never know it - Picasso's body of work is one of the best representations of the Zentangled universe there has ever been or will be.

The closest mental discipline that I can find that is proven over the ages to help in this regard is the philosophy of Zen a form of Buddhism which goes back hundreds of years, a form of meditation that is meant to relax and release the creative energy inside.

When you are in the Zen state of mind, you are focused on Truth and the essence of who you really are, the nature of your existence in this most contemplative state. All kinds of information can flow into your Consciousness when you are in a Zen state. When you are not in that kind of state, information flows but usually in regards to trivial events in your life.

I was in this state of mind when the God Particle was announced by the Scientists using the Large Hadron Collider (LHC) in CERN Switzerland.

They had been working on the problem for years and for several months, word had leaked that they were on the verge of this monumental discovery. I prepared myself by reading everything I could get my hands on about the Particle and the Science behind it, or I should say, underneath it, all around it, in between and all through it.

I will always remember that eventful day for the rest of my life. It was July 4th 2012. My family and I had just landed on Maui for a little vacation when the announcement was made. And over the next few days I started my work on a book about how important this discovery could be.

It was the most enjoyable book I've ever written. I was basically bobbing up and down in the warm rolling waters of Maui, receiving the message from God about the God Particle. So, not only was it announced to the world on that day, but it was announced to me very profoundly and I continued the amplification of the waves of the announcement up until now and probably all the way up until the end of my days.

Using the most complex and the most expensive scientific equipment in history, they had uncovered the greatest force in Nature. They had found the force that creates all other forces. Without the God Particle, things would never be possible to come together the way that they do.

The Large Hadron Collider

WHAT HAPPENS IN THIS instrument is a huge miracle of engineering and technology. The tube you see above actually describes a circle that is about 27 Km in circumference, about 8.5 Km in diameter. It is buried some 300 feet below the ground so that the radiation it gives off cannot harm anything on the surface of the nearby countryside.

Using powerful electromagnets the size of five-story buildings they speed up subatomic particles, mainly protons to nearly the speed of light, with two proton beams traveling around the tube in opposite directions. Then, they suddenly aim the two beams towards each other which has the results of crashing or colliding many millions of particles against each other.

From the tracing of the energies produced by the particle collisions, they are able to infer the most hidden and subtle properties of Nature that are the basis for all of our understanding of the universe.

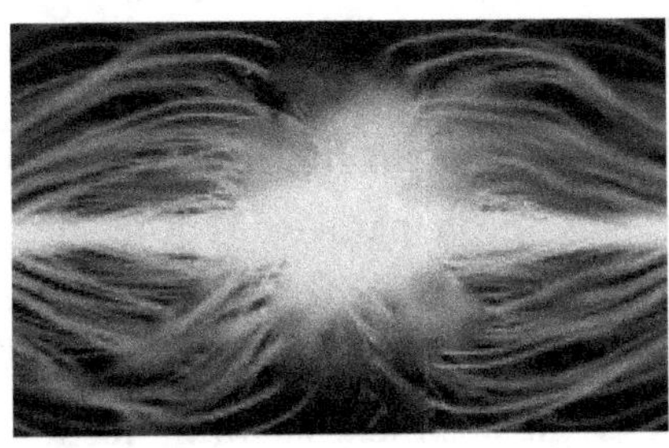

Particle Collision inside the Particle Collider

WHEN WE LANDED AT THE airport on Maui on the 4th of July, 2012, they were making the announcement of this incredible discovery. Even more incredible than the discovery itself, somehow they had focused more than a billion pairs of eyes watching the announcement on TV. That many people were watching an announcement of a discovery in Physics? – Astounding.

This was a first in human History. That many people have not watched the Super Bowl or the World Cup Soccer Match to my knowledge. Nearly one seventh of the population of the Earth, or one in seven people were aware of this moment and actually took the time out of their lives to watch it happen. What happened? It was an announcement of the fact that something that we call the 'God Particle' was discovered. This blew my mind more than the nature of the discovery. And, I would have the first book about it out on the marketplace of ideas. What an honor that would turn out to be. I can't tell you more than that.

Of course, if there were one billion people who actually took time out of their day to watch the announcement, that means that there were probably another billion who could not take time to watch the event unfold, or did not have access to the specific channel that carried the announcement live, but would have watched if they could have done so. That's probably 2 out of every 7 people alive on the planet at the time who cared about Science that much or had some kind of inspiration about it.

I suspect that the popularity had something to do with the fact that there was the word 'God' in the title of the thing. Some people were calling it the Higgs boson, named after the scientist who predicted it some fifty years earlier. But most of the world would understand it correctly using the term for it that depicts its true significance. For most of us civilians out there, it was like the announcement of the discovery of God's whereabouts. And, why not? We finally had his messengers in our cross-hairs.

As I was bobbing up and down in the waves of Maui, I formulated what I would write in each chapter of my new book. When I got up the next morning, I would write out the concepts that I had put forth in the ocean the previous day. So, after our ten day vacation, I had the entire book in the computer. It still needed editing, but I had time to do that when we got home.

This was the happiest ten days of my life. My family never even knew that I was working on a book because I also made time to do things with them, perhaps not as much as most of you would have gotten to do, but it was enough because no one complained. We swam and snorkeled all up and down the Kaanapali Coast. We saw sea turtles and dolphins and whales and the kids were truly inspired, I believe, by what I consider to be the most beautiful and wonder-filled spots on the Earth.

This was my trip to the 'Galapagos' islands.

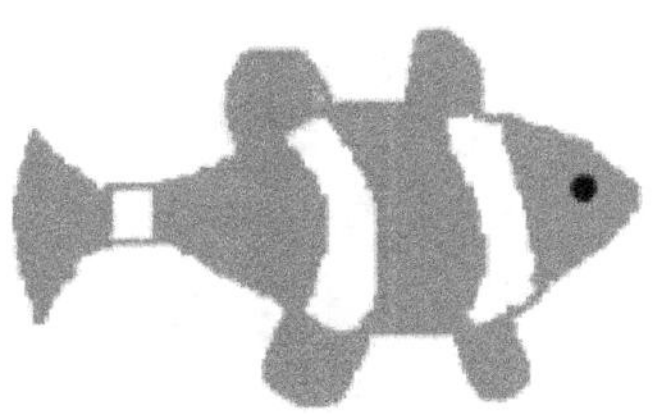

Ideas can also Evolve
From One Fish, Many Fishes

Today, almost a decade since the discovery of the God Particle, I think it's safe to say that God's existence has been proven. I believe I will safely and completely prove that point by the end of this book, if I haven't done so already. My contribution to Science will be how I prove that the God Particle is not the particle that gives every other particle its mass, but instead is the carrier of the "Information" about how, when and where all other particles are to size themselves and place themselves in the universe.

REMEMBER, ALL OF THE information that the universe required was imparted in the first trillionth of a trillionth of the first second of the Big Bang. Ever since my Astrophysicist made me aware of this fact, I wondered, how does all of that information get created and permeated into the universe so quickly. Someone or some 'thing' had to be responsible for this amazing feat. All of that information doesn't get created out of thin air, completely by accident, does it?

So, the rest of my career from this very important day was all about how God performs some of his most amazing miracles. Every life form, every cloud in the sky, every breath you take, every movement you make, every tree, every insect, every bead of sweat is a miracle that pops out of nowhere in the middle of a very hostile and empty Space.

Evolution, the kind you studied in school, 'Survival Of The Fittest' is one kind of evolution, to be sure. However, everything that is being discovered lately by Science is pointing to another higher form of Evolution.

It's the Evolution of Consciousness that comes next, and which has nothing to do with survival in the physical sense. The most important question we must be asking ourselves is: "Are we humans worthy of surviving on this planet as a species?"

Stay with me.

Chapter Two

Entanglements

Is this The Force from Star Wars?

BEFORE WE START TALKING about the latest discoveries in Quantum Physics, I thought it might be useful to first provide a simple demonstration of these concepts to the reader using the laws of physics that everyone who has used their microwave oven to prepare dinner or a snack in seconds already knows.

The microwave oven
Electronic Wave Energy

I THOUGHT IT BEST TO give a little mini-lecture on wave energy for those who have forgotten what this form of energy transmission is all about. When you first turn on your microwave oven, you are turning on the switch on the circuit in your home that allows the flow of electrons from your local electrical power plant down the line into your home and then finally into your appliances. Using this every day appliance, this wave of energy from the generating station to your convenient little invention of the microwave oven, the 'tiny waves' or microwave energy will cook your food from the inside out within seconds instead of what used to take hours.

All microwaves are radio waves, but they are very small and energetic and they are produced by the microwave oven from a very simple little device called a MAGNETRON - seen here.

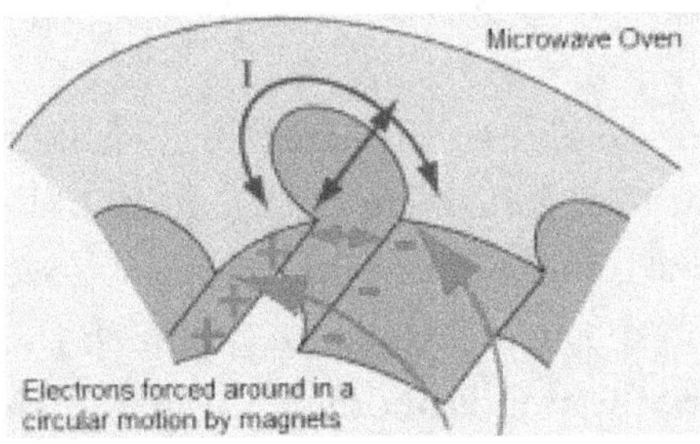

INSIDE THIS LITTLE device, invented in the 1950's by an engineer at Raytheon Electronics lies the heart of the machine. The genius of this little device is that it takes your electrons from the Grid or your Solar Panels and sends them into these round chambers at very high speeds through the use of magnets. When the electrons swim around in these circular chambers they produce the radio waves, you know as microwaves - the blue lines.

The next thing that happens is that your food, sitting on the turntable of the oven, is radiated by the energy of the microwaves. The microwaves are just the right wave length for them to heat up the food in the oven and we all know from experience that the more water in the food, the faster and evenly the food heats up. This is because the water molecule is a very simple one. Most of us can easily remember that water is simply two Hydrogen atoms and one Oxygen atom bonded together and the symbol is therefore - H_2O.

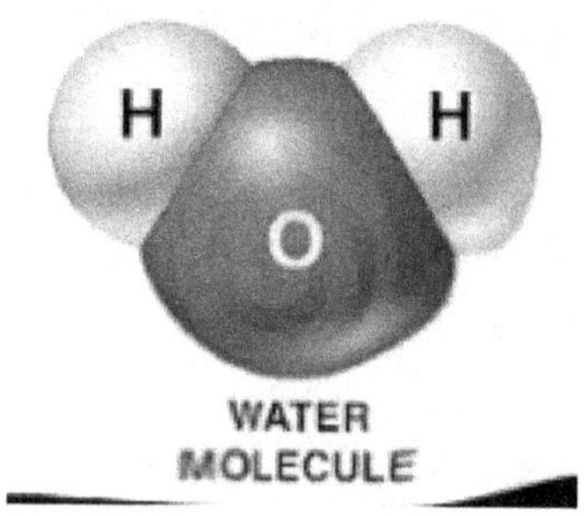

TWO HYDROGEN ATOMS and One Oxygen Atom combine to make water.

So, this molecule is able to resonate along with the frequency of the radio waves so that they gain energy in the form of heat that is created by the incoming wave energy from the microwave oven.

So, electrons in our lives are able to do much to enhance our lives and we are aware of thousands of other examples such as the reading of this book on your computer or other electronic devices.

Now, it's helpful to understand that all electrons are created equal. They have no problems with any kind of hierarchy or predisposed advantages in life as we humans, and all living things do. And, in fact, this subtle property of the physical universe may be where we got the idea and put it into our most fundamental political writings.

As we have seen, every electron, proton, photon, boson, lepton, muon, gluon and quarks are all exactly alike. There are no differences between them, zero, nada, none at all. This is the first miracle of Space/Time in my opinion that there would be a subatomic particle factory somewhere that can spit out unimaginable googolplexes multiplied by googolplexes and that number to the power of infinite googolplexes of subatomic particles being created every millisecond and yet not one of them has ever been discovered to appear to be different or act differently in any way from any of the others.

That may sound odd to you simply because I guarantee you that no other Science teacher in your studies has ever told you that, but it's one of the greatest mysteries in Science, unsolved until now.

But, even stranger, is that every once in a while two of the same kind of particle will become attracted to one another, hook-up in some strange partnership we know nothing about and when you separate them, they long for each other, behave in strange ways and flip their entire personality to mirror their partner whenever their partner is disturbed and changes its personality.

This is known to Science as 'Entanglements'. In Quantum Physics, the latest theory is called 'String Theory' and roughly stated, modern Physicists believe that the smallest elements in the subatomic world that we will eventually discover are no more than strings of energy. They can be long strings, or circular strings like rubber bands. They connect to one another via forces that are yet to be discovered and create the higher energies that we see as the subatomic particles.

In essence, this theory seems correct to me, except that I would rather define these invisible strings of energy as 'Zentanglements' because since they connect everything to everything else in some kind of basic fundamental mesh or lattice, then, this type of energy must be entangled to all other energies of the same species, otherwise, how would they know what to do and in what kind of a lattice or mesh they were supposed to form into? And this may be how everything gets connected to everything else in our universe.

Therefore any kind of String Theory, in my opinion, must include the property of being Entangled for every single string. Imagine a ball of loose strings in your hand. When you pour them out onto the ground, they do not construct anything useful but fall onto the ground in a random pattern making a clump of strings.

Unconnected strings are not likely to form up into anything.
NOW, IMAGINE IF ALL the strings in your hand were tied to at least one other string. Now, when you throw them up in the air or down onto the ground, they are going to attempt to construct something, aren't they?

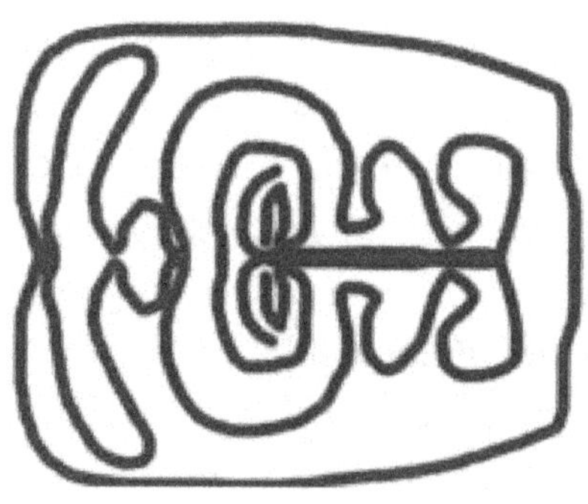

Entangled Strings are far more likely to create something useful out of the chaos.

SO, HOW CAN WE DESCRIBE the Entangled properties in the subatomic world just a little more? Order out of Chaos.

First, we need to understand the latest discoveries about 'Entanglements'. So, we'll use the analogy of what happens when two people find themselves trapped inside an elevator whose size is only big enough for one. In the case of electrons who find themselves in this situation, they squeeze themselves into this impossible space by simply reducing their size in half and maintaining all of their force and direction in what is known as a Half-Spin-Orbit around the nucleus of the atom.

In other words, one electron will lose half of its space and spin UP and the other electron will lose half of its space and spin DOWN. In this condition found in Nature, these two electrons are said to be 'Entangled'. The truly fascinating discovery about Entanglements was first predicted by who else - Albert Einstein, who called it 'Spooky Action at a distance' and even though this was predicted by his own theory of Relativity, he found it very difficult to believe was true and rejected it offhand. Later, he would call it his biggest blunder.

In brief, what Einstein was referring to as 'Spooky action at a distance' is known as 'Entanglements' or when two electrons find themselves trapped inside this tiny elevator and one of them has to spin UP while the other one is forced to spin DOWN. When separated, even over thousands of miles of distance, the two electrons never forget their relationship towards one another and they will mirror each other's spin instantaneously if this state is changed in the other partner. One electron's state is always going to be the opposite of their soul-mate and when one state is FLIPPED by an outside source, such as a human investigator, the other one is FLIPPED to the opposite state, thus preserving forever, the relationship that the two electrons established when first they hooked up. And again, this change takes place instantaneously, regardless of the distance between them in total disregard for the element of Time. Time does not exist in this particular feat of Nature.

One of the proofs came in 1964, when physicist John Bell proposed that quantum entanglement could be demonstrated by separating the particles at a great enough distance that any correlating effect on both particles could not possibly be caused by local environmental factors. These were called the Bell Test experiments.

When separated, he noticed that when he changed the spin momentum of the UP electron, the other electron in the PAIR - what I will call the 'Soul Mate' of the other electron somehow senses the SPIN CHANGE and IMMEDIATELY REVERSES ITS OWN SPIN to counteract the other partner.

They do this in Zero time differences, no matter how far the distance between them, thus defying all known laws of Physics. This is the 'Spooky Action', in Einstein's remarks.

The first question to be asked, and which is as yet unanswered until now, is 'How do they know?' How do they know about each other and how do they sense their partner's spin state? They have to 'know' it somehow, yet electrons and all other subatomic particles have no brains, no eyes, no ears, no sensory organs of any kind as far as we know.

But, when they all immediately react to the fact that their soulmate has changed it's spin state, something is obviously going on between them. To know something usually requires a storage of knowledge and memory. They don't have this either. So, how do two entangled particles learn about the condition of their spouse? Somehow, there is an exchange of this information at some unseen level.

I call this the first evidence of Consciousness. Don't worry, we are going to dive deeper into that aspect of this story in the next chapter. For now, it's only important to know that these subatomic creatures are doing things that we would ordinarily put down to living, intelligent beings. Yet, there is no evidence that anything like a brain or neural system can exist at this level of Evolution.

And, the second question - even more important than the first - is HOW do they perform their SPIN REVERSAL - in total defiance of all known laws of physics? Where do they get the energy to stop and reverse their spin. In High School we all learned about momentum and the First Law of Physics which states that all matter that is moving in one direction will tend to remain in that motion until acted upon by another force.

Got you stumped already?

Don't worry, by the time you complete this book, you will have the answers to what I believe to be the most amazing and potentially rewarding questions in the history of Science.

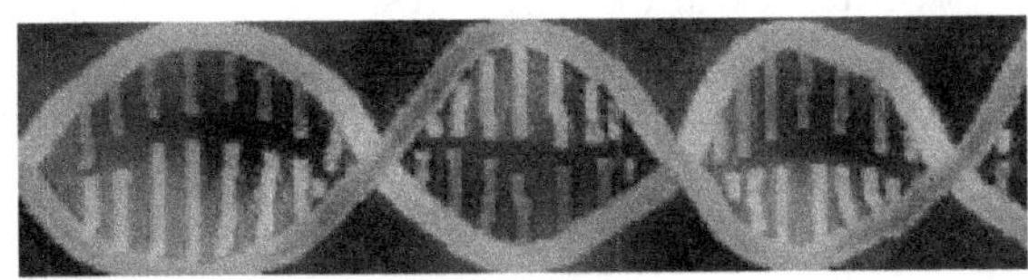

THE IMAGE ABOVE IS what I perceive as the basic shape of our DNA molecule inspired by our previous discussion a few moments ago about String Theory. With this basic molecule every form of life on planet Earth is derived. Every cell in your body is composed of the same DNA molecule as composes every cell in every horse, dog, cat, chimpanzee, orchid flowers, the leaves of all the trees, all the blades of grass, all the insects, all the fish that swim in the seas, and even including every other life form all the way down to the lowliest bacterial form of life.

In fact, I have predicted in an earlier work of mine, 'The Origin Of Creation' that every other form of life that we may someday encounter in our space travels will also be composed of this same DNA molecule. The programming of the base-pair nucleotides shown as the stair steps in the diagram above, will produce a completely different set of molecular codes, genes, for different forms of life that have adapted to another planet's living conditions. So, life on other planets can still appear extremely alien to us.

But, though the coding may be different, I predict the molecule that contains all of the code, enough to code for all life in the universe will look and behave exactly as the DNA molecule behaves here on Earth.

Another thing to notice about the basic form of the DNA molecule is that it appears exactly like two strings wrapped up around each other and influencing each other into forming the Helical shape that goes on for billions of base pairs.

Could the physical knowledge of each other in the Entangled state of the subatomic universe be the cause for the way that DNA is constructed? I wonder. I want to say – yes.

In a thought experiment of mine, I performed several years ago that produced the book – 'Maxtricity – The Science of Entangled Electrons', I reasoned that if you connected a long string of 'Entangled' electrons you would have the following structure. Think of this as a string of electrons that instead of flowing along a copper wire are merely sitting next to one another on the wire, not moving, but instead spinning in opposite directions to their neighbor. In opposing one another they would create a clock-work set of 'Gears' with one electron pushing the other into the opposite motion, thus producing what I call 'Maxtricity'.

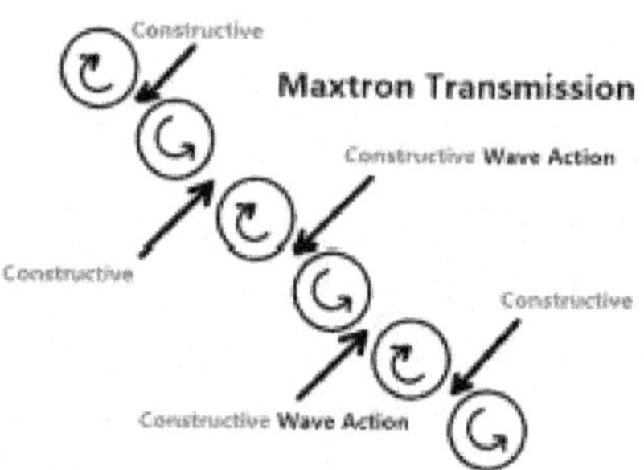

I call these Entangled electrons Maxtrons because you get the Maximum Electrical power out of them

IN OTHER WORDS, IN Maxtricity, electrons are not flowing along a wire, as we have them do now whenever we use the force of electricity. Instead, the Maxtrons (Entangled electrons that achieve their maximum power) remain in place, but spin and then create an opposing spin with a super-conductivity that helps make this almost frictionless transmission of power. Of course, if accomplished, this would solve this planet's energy crisis for all time.

Think of it like a transmission in your car that does not require an engine. The gears would simply drive themselves to higher and faster outputs. Unlimited horsepower without the need for gasoline or batteries.

Today, in hindsight, I don't see this model of mine actually working because it means that all of the Entangled electrons are going to influence the next one in line. However, that would require the entangled electrons to be entangled with a partner on their left for one moment and then entangled with their partner on their right in the next. Perhaps some genius electrical engineer in the far off future can figure out how to do that, but to me it's an impossible feat, at least for now.

I only mentioned this little pipe-dream of mine so that more of you might be able to imagine the power of these ideas. Real benefits may someday come out of them.

The main thing I am hoping you remember about 'Entanglements' at the subatomic level is that these emotional ties that particles make towards one another, are very likely to be the basis of emotional ties that we make with one another. Where else does love, fondness, caring, longing for each other come from?

Yes, some will say love and affection are merely biological functions, the workings of our hormones and gonads working on our brains to make us yearn for one another, the pressure to procreate, thus preserving the species in future generations. And on a certain level, it's true. This is how things work technically speaking in our bodies. But, where did the nuts and bolts of our biological selves get the idea to perform these tasks of making us yearn this way? In my humble opinion, they got them from the subatomic level that forms all things, even the things that inform our instincts. There is no other explanation.

BEFORE WE LEAVE THIS chapter, I must mention something just as wonderful as everything you've learned so far, and perhaps even more wonderful. And it fits our discussion in a way that should stimulate your brain to arrive at some of your own answers.

Something also recently discovered is known as the . . .

'Zinc Spark'.

If you google - Zinc Spark, you will find evidence uncovered by Thomas V. O'Halloran and Theresa Woodruff about a recent discovery made by watching and closely filming the moment when a sperm cell enters the egg cell, or the moment of conception of every living mammal, including you and me.

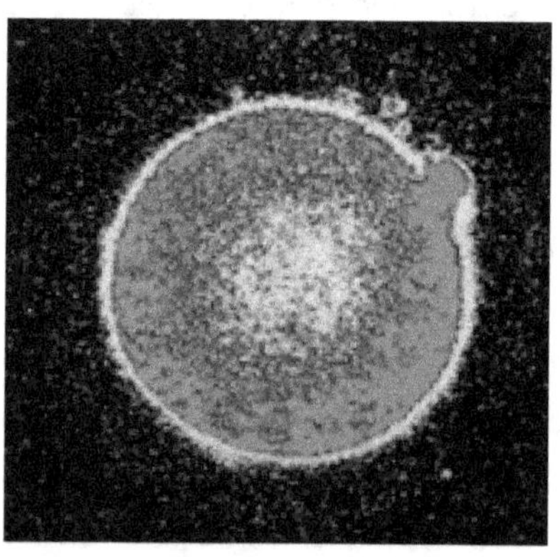

First, there's the red color of ignition.

THEN THERE IS AN EXPLOSION of light created by about 20 billion zinc atoms.

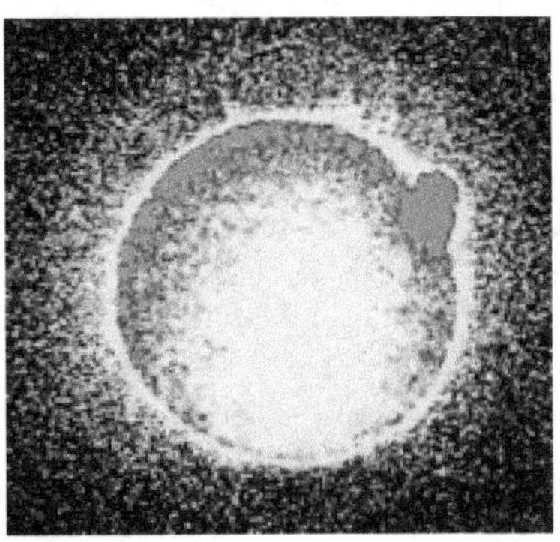

These are pictures taken by the scientists who discovered this amazing phenomenon at the point of our birth, the fertilization of a human egg by a human Sperm.

AT THIS PRECISE MOMENT of successful fertilization, a living being is created. Zinc atoms fluctuate in and around the egg cell and do so in such a fashion that they give off light. In essence they create a massive light show, a fireworks that announces the creation of every creature on Earth classified as mammals.

Not only does this amazing fireworks display announce each and every one of us at the moment of conception, but the fireworks show continues in waves into the galaxies of cells that are rapidly dividing and being created from the original egg cell in the process we call mitosis. Without this process, we will not turn into a human being and will not be born.

At this early stage of mitosis, it is the Zinc atom that helps the cell division process in such a way that each cell learns what role it will play in the body. Some of your early cells will develop into heart muscle and begin pumping blood around to the other cells. Other cells will learn how to become bone tissue to support the other tissue in your body. Other cells will learn how to become brain cells, forming what will become your personality and the seat of your consciousness. Other cells will learn how to become parts of your fingers and toes. Other cells will learn how to develop into eye cells that allow you to see, ear cells that will allow you to listen.

This is also the point when your chromosomes from the DNA of your mother and the DNA of your father come together to create your own copy of DNA. You are physically 50% your father's DNA and 50% of your mother's DNA. This is true of all life forms on the Earth, since all life is constructed out of DNA.

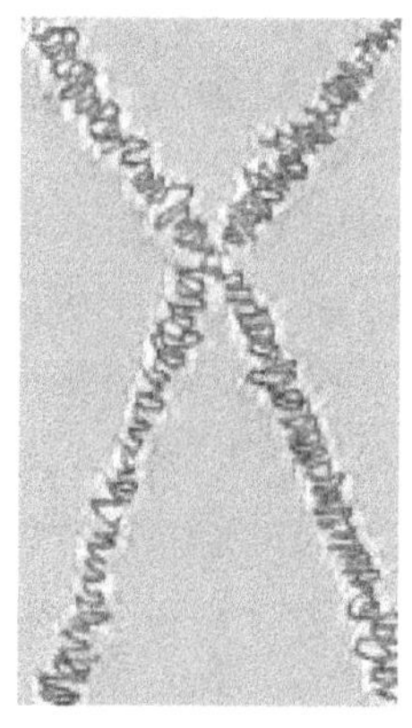

THE TWO HALVES OF YOUR parents DNA combine to create you also at the time of the Zinc Spark. It's one of the greatest and most amazing discoveries in my life to learn that there is such a beautiful light show that announces every one of our moments of creation. To my way of thinking, life is so rare and so wonderful and so magical to our Creator that he, she or it marks a special occasion that is witnessed in this special way even though it means that there are billions of these every second.

All of the development of who you are and who we all are as a species is dependent on the element of Zinc to transmit this information throughout all of your early development in the womb.

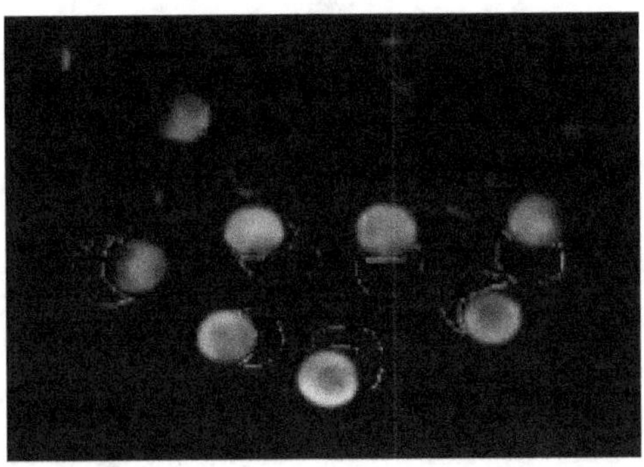

Early cells undergoing the 'Zinc wave'.

IN EVERY SENSE OF 'THE Word' as given to us by modern Science, if it weren't for the mysterious side of the universe, the zone where there are more questions than answers, we humans could not exist, or if we existed, we would be as the millions of other species on the Earth who have no words nor any language or any other higher logic system for deciphering the world around us.

The fact that we can look up into the night sky and recognize all of the gravitational movement in the universe and down into the darkness of the smallest quanta of energies known and translate their movements and properties into the libraries of information that we know as 'The Truth' is the most wonderful of all discoveries that we have made or will ever make.

Is the final Truth yet to be revealed?

Let's find out together.

Chapter Three

- The 3 Laws Of Consciousness -

I must tell you first that I am not a Scientist in that I am not now nor have ever been employed by an educational or scientific organization. This has never been part of my career path. Instead, I have been an independent thinker all of my life as far back as I can remember. But in my independent research, I feel that I have been freer to investigate the questions that truly need to be answered, rather than minor insignificant ones. And the biggest question I've had throughout my life is how events in my life are ordered, by whom are they so ordered and to what purpose are they so ordered? Are these indications of some kind of universal laws? I think that you might agree by the time you finish this book that I may have some answers about that.

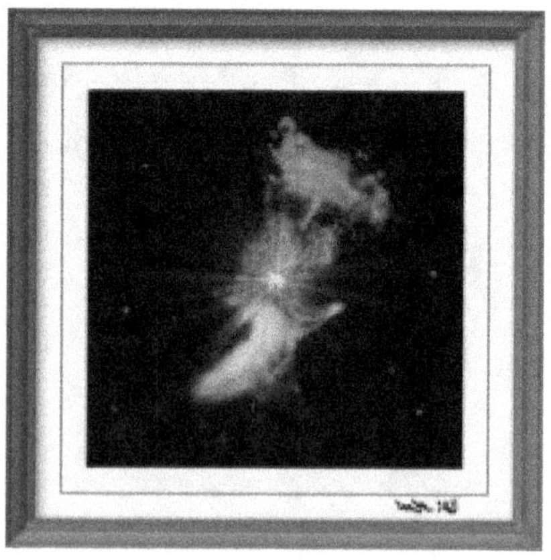

The 'Hand Of God' Nebula - pulsar B1509-58.

DON'T WORRY, THIS IS not 'Rocket Science'. These are just basic mental agility exercises that present no real challenges to anyone capable of reading the words. If you feel any anxiety or discomfort in going further, please just put the book down, with a bookmark, in this page so that you can renew your reading later.

Above anything else - DO NOT - give up completely at this point. You are very, very close to the end where it all snaps together. This could be the greatest moment in your life and in the life of all humanity, if we can all get it together at the same time. This is the Zentanglement of this book. Just grok that one little feature of it, and the rest is a piece of cake - OK?

In all of my writings, I always attempt to reduce things down to the simplest of terms so that I can understand and uncover the larger perspectives and present them to my readers in the highest yet simplest form. And, as you have seen, I like to use illustrations - and now 'Zentangled' illustrations. But, we've just been warming up. These new ones will be coming at you at a breakneck speed now.

To put an exclamation mark (!) on this type of idea, I am inserting here my favorite photograph and what I believe is the most important photograph ever taken.

Photograph taken by Voyager as it left the Solar System. Courtesy to NASA/JPL

IT'S TAKEN FROM OUR Space probe Voyager in 2012 as it left our Solar System, the first man-made object to ever do so. I wonder how many of you can tell what this is all about.

As the SpaceCraft Voyager left our Solar System, Mission Director Carl Sagan had a signal sent to the spacecraft to turn around and snap one last photograph. Not knowing what would come out of it, the NASA team waited several hours for the data that would be assembled into this picture above.

At first, they thought that the tiny white spec you see in the yellow beam of light in the Top-Middle of the picture was a grain of sand or an imperfection in the development of the photo. One of the NASA people tried to wipe it off the page. It wouldn't budge. So, upon closer examination, it became clear that the 'little blue dot', as Carl Sagan defined it, in the Top-Middle of the picture is our own planet Earth, the originating point in Space from whence the little spacecraft was launched thirty-five years previously.

The finger of orangey-yellowish light that the Earth is floating in is merely that, a finger of light emanating from the Sun, about 90 million miles to the leftward orientation of this picture and very graciously pointing out for everyone to see just how small and insignificant a place our planet takes in support of all our words, deeds and actions.

A person can look at this photo very coldly and objectively and see only what is shown and make no major interpretations of this event other than the astounding fact that the equipment on board Voyager was still working in this completely hostile area in space and after so many years past everyone's life expectancy of the intricate hardware involved.

Or a person can learn that this picture was never planned in all of the thousands of hours of designing the space craft's mission. They would have their hands full orienting the spacecraft to snap millions of photos of Mars, Jupiter, Saturn, the moons of Saturn, Uranus and Pluto as it flew past all our immediate neighbors at about fifty-thousand miles per hour. No one ever thought to turn the cameras back for one final shot from this distance where they knew that nothing would be there worthy enough to photograph, or so they thought.

In fact, I believe it was the suggestion of one of Carl Sagan's assistants to send that one last command to the spaceship to turn around and take a random photograph like this. The brass at NASA had rejected the idea thinking that nothing could come of a very expensive, multi-million dollar maneuver. And besides, they told Director Sagan, the project had been shut-down. Everyone had gone home. It would take eight years and six separate requests for the brass to agree to the last order.

Carl Sagan persisted however, because it just seemed like such a cool idea and eventually he was able to convince his bosses to call several project engineers back to their command offices at Jet Propulsion Labs and get to work to send one last signal to Voyager.

And so it was done. It amazed everyone at JPL that Voyager still had enough juice in its batteries to not only listen to their signal but also to obey and turn itself around, focus it's camera in the exact opposite orientation to where it was going and snap one last photo and transmit the data back to the Earth.

Whenever I assess all of these points of data, I come to the only conclusion that someone like myself, and I hope you too can make.

The little Voyager space-craft had one last job to do. It somehow made this known to its creators. They took up the challenge and allowed Voyager to turn around and wave good-bye to us, the creatures who had made it, and in doing so, Voyager would give thanks and memorialize the event by pointing out to us exactly how rare and precious an element we are in the overall scheme of things.

Now, I have to say it - Compare this actual event that is abundantly recorded in digital form that will last for millions of years to the story of Abraham when, allegedly, God ordered him to kill his only son. Which of these events would you trust as proof of a God? I mean, can you really picture our God and Creator ordering one of us to kill his only son? Come on, man. (Remember, no Dummies are allowed here.)

And if the Human Race goes extinct someday and no other living beings are able to contact us, they may someday bump into this little space craft that we constructed and now plying the depths of interstellar space. They will take it apart and learn how to play the music and videos that we placed on a golden record filled with images of our own lives and many of the life forms whom we are honored to share this planet with. They will hear music from Bach, Beethoven, the Beatles and Chuck Berry. They will see images of babies, flowers, exotic fish, animals running wild in the jungle. High above the savanna, Eagles and other majestic birds float majestically in the sky. They will even be shown a map of how to find our planet.

When they arrive, they may find that we have destroyed everything here and that there are no wondrous animals and plants adorning this Earth any longer. There may be nothing but another burned out and lifeless planet left for them to see.

And if they are good creatures, they will mourn for us and maybe even feel deeply sorry for us. They will at least, from the records and data on board Voyager, know what we looked like, what we have striven for most arduously, the kinds of things that made us happy and even how thoroughly we loved each other.

But, they may never be able to fully understand how we lacked the intelligence and the forthrightness to save ourselves.

A Quark Star
Voyager Photo – Zentangled. Voyager turned around in its flight away
from the Earth, over nearly two decades to look back at it at the
precise moment when sunlight pointed it out.

Voyager Photo - on Canvas

THE REASON THAT I POINT to this photograph from Voyager
as it left our Solar System is because the series of extraordinary events
actually evidence the Consciousness of the emptiness of Space that I've
been describing and that Einstein described as a fabric.

Don't worry - I have the mathematical proof of my theory coming up later in this chapter. However, before I get to the cold, objective and mathematical proof of this incredible foundation of our universe, I want as many of my readers as possible to appreciate the events that have taken place in all of our lives intended to expand our consciousness in a very big way. The Voyager mission is one of my favorites, but I have many more to show you.

BEFORE WE GET TO THE final piece of the puzzle, I need you to consider one more recent major discovery in Science that millions of people have learned about but fewer than that truly appreciate the meaning of it.

LIGO is an acronym for the Large Interferometer Gravity- Wave Observatory. It is able to detect waves as small as one Ten-Thousandth of the diameter of a Proton.

This is the most sensitive instrument for observation of the heavens ever created and the minute they turned it on in 2015, it recorded gravity waves, as it was intended to do, from 1.3 billion light years away. This instrument is the opposite kind of instrument as the Large Hadron Collider which is the most sensitive and expensive scientific instrument for observing the smallest things in the universe. Isn't it a fun fact, that we as humble life-forms have developed both of them and at about the same time frame?

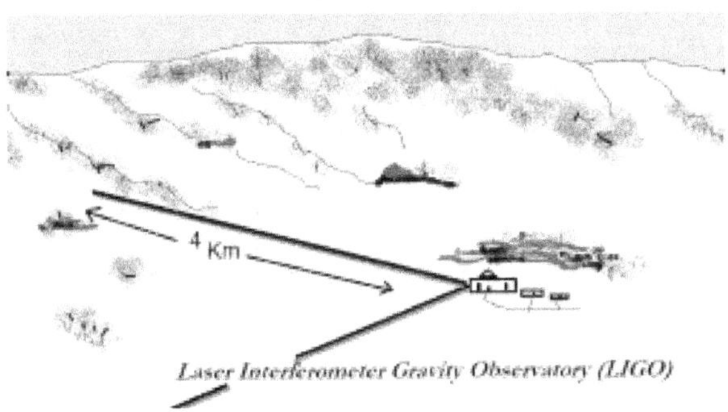

Laser Interferometer Gravity Observatory (LIGO)

REMEMBER ALL OF SPACE/Time is a kind of fabric. So, Scientists who designed LIGO reasoned that the fabric of Space/Time should be bent or warped by a very small but measurable amount whenever a Gravity wave travelled over it.

And, so LIGO was designed to pick up on that wave as it sped past the Earth. But, they never considered how the first one would be uncovered.

When they were finally able to turn on LIGO in Sept. of 2015, they had been working day and night for months. Exhausted, they went home to go to bed believing that the chance of the equipment picking up a Gravity Wave on the first day of operation would be one in a billion.

They were wrong. As soon as they got home, they received an alert from the station. It had picked up on something significant. They sped back to the station and saw the data that was telling them that on the night of turning this most sensitive instrument in history into operation, it had in fact uncovered a gravity wave.

LIGO is constructed in identical systems in two places in the United States. One in Louisiana and the other one in Washington State. The reasoning for this is for the prevention of them recording something like an Earthquake or another local event that causes the equipment to record something. A Gravity Wave from outer Space would obviously have to hit one area on the planet and then travel onwards past a second location at the speed of light, thus lending credence to the event of the capture of a Gravity wave if captured by these two locations several thousand miles apart.

And, that's what happened. But, just to be on the safe side, the scientists working at LIGO kept the discovery from the world for 6 months as they constantly reviewed the data because even they were convinced this had to be someone hacking their system or a problem in their equipment. They found it astounding that the world's first Gravity wave discovery would happen on the opening day of their equipment.

After six months of reviewing their data by thousands of colleagues all over the world, they finally had to share this unique discovery with the world and did so on Sept. 15, 2015.

But wait – there's more. The actual Gravity wave that they found took place over 1.3 billion years ago, or 1.3 billion light years away from the Earth which they calculated that this kind of event could only have reached the Earth if it was caused by two Black Holes colliding with one another.

Sure enough, when they looked up in the sky in the direction of where they knew the Gravity wave had to originate, they saw the visual evidence of two black holes colliding.

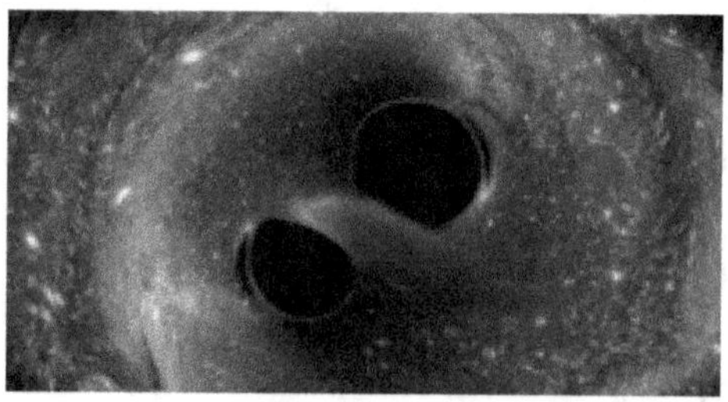

Two Black Holes Colliding

AND, SINCE THIS INITIAL discovery, the same kind of thing has been detected by other LIGO experiments that have sprung up all over the world.

I mentioned the 'Fun Part'. So, this first gravity wave that we detected, proving the nature of the so-called emptiness of Space is a kind of fabric, took place 1.3 billion years ago. But, we detected it on the first day that we have the ability to detect such events.

And here's something that should truly blow your mind. All life on this planet is barely 1.3 billion years old.

Coincidence? Perhaps, but hold that thought.

But wait - there's still more. When the scientists at LIGO translated the gravity wave they detected into a sound file, it sounded like a 'Whistle'. You can Google it and hear it for yourself.

So, over billions of years, someone or something alerts us to the nature of their universe by sending us a Whistle so that we don't miss another important clue to finding them. But my dear reader, we hadn't even been born yet. However, it must have known that we would be born and matured enough to understand it by the time the signal got here.

'Hey, you Bozo's down there, here's a little birthday present for you,' it seems to whistle.

Now, many people can go over these events and claim even stranger conclusions for you. But, this is the only source that can prove what they are saying.

The Mathematical Proof:

First, I need you to recall the Standard Model of Physics from the first chapter.

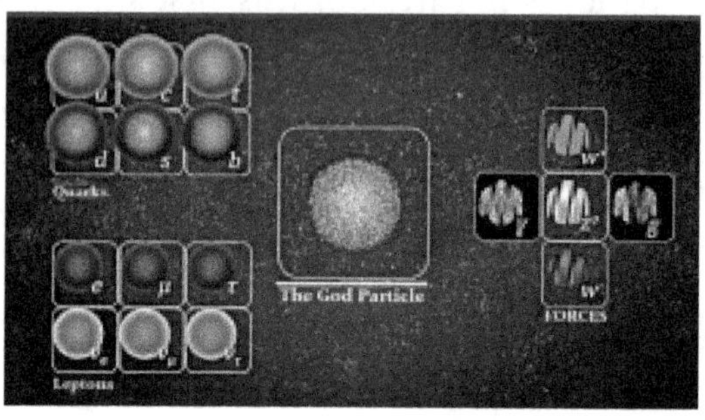

Everything in the universe is constructed from these 18 smallest of building blocks and nothing else that is known to us.

WE ALSO KNOW THESE so-called 'particles' of constructing the universe are really waveforms that spin around so fast in one place they can appear to us to be solid particles, but they are not and never will be anything that you could put in a spoon to stir in your coffee or or into a sling-shot and shoot at someone.

The laws of Physics that we teach in Universities all over the world are based on the discovery of these kinds of subatomic energies and over the years we have found more and more of them. No doubt that someday, we will find more still and so our Laws of Physics will be forced to change to accomodate more and varying forms of energies at this level of reality.

But, the Laws of Consciousness will never change because it is this basic fabric of Space/Time that keeps everything on the same plane. And even though that plane may be multidimensional, it makes no difference.

One of the other great things we teach besides basic Laws of Physics in the Universities is also the science of Mathematics.

One of the basic cornerstones of modern Math is the number, or lack of a number, Zero. We use 'Zero' all the time. If you are using a computer to read these words or listen to them, the device you're using is using the 'Binary' system of math, which means you don't count higher than one. Everything is calculated by 'ones' and 'zeroes'. There are no other numbers inside a computer's brain, even though it can calculate all other numbers and so your computer, your smart phone, works for you the way you want it to.

But, Zero was not used in math until three centuries after Christ. Even after that, it was made illegal in some countries for religious beliefs. They thought that there couldn't be anything of zero value because 'God' was part of everything. They were right, in a way. It's just that their concept of 'God' was quite limited. As it should be given the extremely minor amounts of Science that had been compiled at the time.

In recent times Stephen Hawkings, one of the great geniuses who many believe was in the class of Albert Einstein, said that the entire universe was a 'Zero sum game', meaning that everything in the universe comes around in the end to nothingness or zero. He too, was right in a way, but he possessed a very limited concept of 'Zero' and God was non-existent.

The most important aspect of this number, however, is that Zero is now the basis of all modern mathematics, including the math that takes place inside our smartest machines.

Simply put, any professor of Mathematics will show you today how we must allow for any number that is raised TO THE POWER of Zero to equal 'ONE'. It turns out that 'Zero' is the most important number or lack of a number of all of them.

This is how I came to understand my 3 Laws of Consciousness. To demonstrate the process in my mind you must follow my logic which took place over a couple years.

Take this number, a very large one, for example.

34,000,000,000,000,000,000,000,000,000,000,000 and raise it to the power of Zero . . . which would be better shown in scientific notation, but for simplicity I'll show it this way . . .

$$34{,}000{,}000{,}000{,}000{,}000{,}000{,}000{,}000{,}000{,}000{,}000^{(0)} = 1$$

Now, I can take this number or any other number and add an infinite number of zeroes to it, raise that number to the power of zero and it will still equal 'ONE'.

The same goes for negative numbers. In other words, the largest things in the universe, or the smallest things in the universe, raised to the power of zero will always equal 'ONE'. Don't believe me? Well just make up the largest number you can think of using the largest chalkboard you have, make it a positive number or stick a '-' minus sign in front of it and put all of the zeroes right after the minus sign to make it the smallest number you can think of. Then, put a simple (0) next to it as its exponent and those two numbers, or as many as you like will all total to the same number 'One'.

This is probably why Stephen Hawkings said that the universe was a Zero Sum Game'. Because, mathematically, it is.

But for me, the mathematical breakthrough was seeing this very clear connection between Zero and Infinity.

This gave me the idea to make a corollary of this basic fundamental mathematical principle. When you realize that any number how small or how big, raised to the power of Zero equals '1', you must agree that there is a distinct correlation between Zero and the infinite size of the universe, no matter how big or how small.

Now, it's absurd to think that all of these numbers, any number in the universe no matter how big or how small upon interacting with the power of Zero can equal the number '1'.

It's just not rational to believe it. There is no way that can happen in the common everyday universe we inhabit. So, what the formula is actually indicating to me is that any number raised to the power of Zero can equal 'Unity' or 'One-Ness'. In other words, they all have their most basic fundamental purpose in common with all other numbers when they are influenced by the power of Zero. They become the same thing. They are interchangeable. This is rational. I can understand and agree with that principle.

For you to also agree, you may need to remember that numbers are really symbols of a mathematical entity. In different languages, numbers, as symbols may look completely different, but the placeholder that symbol stands for in any language is always the same.

Put another way, this formula is stating something that is almost instinctive in our minds which is that a fundamental and mathematical relationship that we can delineate between all numbers. And, in this formula, it is the uniformity that we see when we look out at the infinite distances in Space, or within the smallest subatomic distances and notice that everything works exactly in the same way everywhere. This is the uniformity that the formula points to. Any number can be 'One' with any other number. They only need to share the power of Zero.

This is also proof of the 'One-ness' that we feel sometimes when we are at our highest and best purposes and we say that we are 'At One with a certain person', or 'At One with a certain place', or 'At 'One with the universe.'

So, knowing the true nature of this formula, we can then do a small reduction in terms, solving for Zero and my own formula will emerge in your minds.

Since any number, through an Infinite power of Zero, can magically become 'ONE', then it is also true to say that any number is equal to Zero to the power of Infinity.

And, so first I had to conceive of a new number - Zero to the power of Infinity.

FROM HERE, IT'S A SMALL jump to this new number being equal to any number, no matter how big or how small. It's probably the simplest math you've ever done. It's even less complicated than one and one equals two, if you think about it – because we're never told why two ones can equal two. We just accept it and go from there.

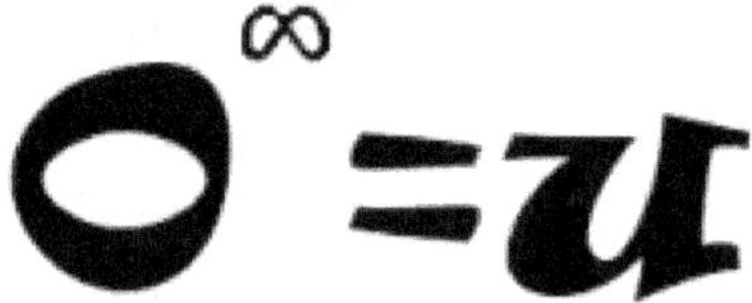

BUT, NOW YOU KNOW, from my formula that any number – U – can equal Zero to the power of Infinity. This means that ALL NUMBERS are equal to Zero to the power of Infinity. All numbers are part of the fabric of Space and Time. The fabric is the 'One-ness'.

Zero to the power of Infinity can equal any number in the universe. Because if any infinite number can be raised by the power of Zero and that new number always equals '1', then Zero raised to the power of Infinity can equal any number. I hope you can see this. It's merely a reverse logic substantiating the connection between Zero and the Infinite.

It's also a great way and also the only way to keep all the books of the universe balanced in terms of a 'Zero Sum Game' as Stephen Hawkings has suggested.

In my equation 'U' stands for any number or it can stand for any place in the Universe because when you think about it, any place in the universe is simply a set of coordinates. Think of this in the same way that the GPS system we have in place by satellites orbiting the Earth can pinpoint you or any place on this planet by a set of numbers, so can the Cosmic Consciousness know where everything else is in the universe as long as places in the universe can be defined by mathematical coordinates. Zero to the power of Infinity can know or physically be in all of those places at the same time. And all without the use of any satellites either.

Until my formula was created, many of us humans understood this concept instinctively. God is everywhere. We all know this, or at least some of us know this - But, now we have our proof in this formula but also in the latest observations in Astronomy. On the one hand, we now know that the universe is all held together by gravity, but we also now know that there is another force equal to or greater than gravity that is forcing the universe to get bigger and bigger and even faster and faster. We call this force – Dark Energy, only because we don't have a clue where it comes from or anything more about it, except it appears to be the opposite of Gravity.

Now, I'll give you a clue about the rest of my line of reasoning. It occurred to me that if there is one force – Gravity pulling everything closer together and there is another force of equal or greater force that is pushing everything apart, then that is the greatest magic trick of all. Space/Time is actually drifting out into greater and greater distances away from us and at the same time being held together by what we call Gravity. This fabric of Space and Time is made out of ZERO stuff. It actually has no distance, no physical aspect or dimension of any kind – only zero size, zero distance, zero energy.

In this way, Consciousness has no boundaries. It is not big. It is not small. Consciousness cannot be measured because it is really a thing like Zero, but raised to the power of Infinity. Therefore, Consciousness can be anywhere in the universe at the same time. It can even be in a multitude of places in the universe because there really is zero distance between all places in the universe even though those distances may appear to be infinitely large when we observe the outer parts of the universe or infinitely small when we observe the inner parts of the universe. But this is because we must always observe the universe through our senses, or through the telescopes or microscopes, our technological sensors, themselves amplified versions of our biological senses, which are easily fooled.

It is all the same to the mind of God and it was and always will be intended to remain that way.

It's also fun to note that the Zero was not widely used on this planet until merchants over one thousand years ago found that it could be used to balance their books. Until that time, they had to use an Abacus to do their bookkeeping which must have been a nightmare for most of them.

But, when they learned about the invention of the Zero, they were able to finally balance their books which meant they were more likely to be successful, catch the cheaters in their midst, find a better way to price their goods etc. The entire modern economic system of Capitalism would not, could not work at all without the zero.

And so it is that with my invention of the new number Zero to the power of Infinity or $0^{(i)}$ Physicists, Astrophysicists, Quantum Physicists and Cosmologists now have a way to balance all of the known and soon-to-be-known energies in the universe so that the sum of all forces known come out to Zero, which to me is also the answer to the question Einstein had regarding the Unified Field Theory.

And with this background information and more that you will read about in the next chapter, I devised the ...

Three Laws of Consciousness:

- Consciousness creates all things
- Consciousness connects all things
- Consciousness consists of all things

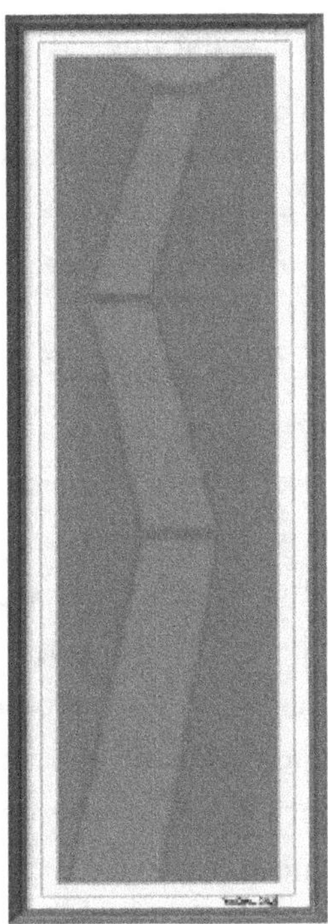

Finally, I want to wrap everything up in the context of current events and 'Zentanglements' as the true meaning of the word that I have coined.

As I author this book, the Corona Virus Pandemic is raging for the second year in a row. There's talk that the virus may have been created in a military biological weapons research lab in Wuhan, China. I don't know if that's true. I doubt anyone will learn the truth about the origin of this terrible killer virus because those in power are very good at hiding the truth from us all until we reach the point where we can't do anything about it.

So, my solution is for everyone on the planet to get ahead of the truth. Because if an intelligent species like the human Race is supposed to be can't learn how to see a little bit into the future and do things proactively to avert total disaster, like the Dinosaurs, we will assuredly go extinct like the Dinosaurs. Of this, there can be little doubt. Because of this current crisis, the existential dangers of allowing nuclear bomb technology to spread around the world including into terrorist fanatic groups, the continual pouring of carbon dioxide into the atmosphere, the poisoning of the oceans, etc. there should be all kinds of alarm bells going off in all of our brains, forcing each one of us to do something every day that will help avert total disaster.

I have an earlier work, 'Extinction Live', a work of Science Fiction, that predicts the end of the world by the end of this century and all I did was follow the logical progression of current events to create my story. I am scared to death that my story will be an apocryphal one.

But, the sad fact is, I don't see the dedication of every individual to do something to avert the obvious disaster. Yes, there are a few thousands of people who make a career of raising consciousness on this issue, even children doing so, and God Bless them all, but as soon as they appear on our radar, they are replaced with news of a senseless murder somewhere or a senseless act of unkindness somewhere, or the politician's latest speech, or a useless piece of news about a celebrity pregnancy or idiotic unimportant events of that sort.

And so, this book may be my last one as I am getting older and I'm slowly becoming convinced, seeing signs that the world may need a huge flushing out of the evil that has accumulated of late and then the whole system restarted millions of years from now when all of our DNA programming is shoved down into the dust and a new resorting of all our genes can begin.

But, I offer here one last weapon in the fight for Truth and wisdom and the ability for the average citizen of the Earth to know what to do about things, how to reorder their lives so that they have the greatest impact on our common survival and maybe even including a little genetic improvement along the way, enough at least to ensure these kinds of things never happen again.

It's the use of my 'Zentanglements'. Improving one's Consciousness is too much to ask of the average person. Even as I have described it here and made it as simple as I can, I know there will still be billions remaining on the planet who just don't get it or will never want to get it in time.

So, I am inviting everyone within the sound of my voice, anyone reading these words, to find at least one of my diagrams in this book and make it a personal endeavor to study it and copy it as many times as you need to have it ingrained in your brain.

They are all listed in the back of the book in the chronological order that they were included in the book. The advanced student of my concepts should be able to scan these images over the course of a few minutes, contemplate them at an extremely relaxed moment in the mind, and reach a deeper level of understanding of my concepts than even I have - assuming I have designed the images properly. In fact, if I have designed these 'Zentanglements' properly,

And, the Truth is coming to us through the attack of the CoronaVirus. If you have been paying attention, you have probably gotten a shot in the arm of a Moderna, Johnson & Johnson or Pfizer vaccine to prevent you from getting the virus or at least getting it badly enough to kill you.

The most interesting thing about these new vaccines is that they are manufactured using the MRNA (Messenger RNA) of the virus itself. First scientists had to isolate the DNA of the Covid virus. Next, they were able to find the Messenger RNA of the Virus and from that code were able to create the Vaccine the way to defeat it.

Our modern vaccines use the Messenger RNA to create an artificial spike protein that is similar to the Covid Spiked Protein, but there is no Covid virus actually introduced into the body. However, because the Spike Protein resembles the Covid Virus Protein almost exactly, your immune system suddenly recognizes it and can fight it if and when the actual Covid Virus invades your system.

In order for there to be a Messenger, there must first be a Message. This is where these highly interesting discoveries of Science come into our body of Truth at the precise time that we are all able to grasp it and use it. It's literally a battle of wits. But, you have to ask yourself - how does a Virus have 'wits' of any kind? These are one-celled life-forms. There is no brain to think with in a virus, is there?

Don't concern yourself with invaders from Outer Space or even from other countries because we are currently fighting a war of survival between the human animal and the survival of one of the first forms of life on this planet - the virus.

We are literally at the point of asking ourselves - who is smarter? The smallest of all living things or us - creatures who have evolved millions of years beyond them?

And so, here's the wonderful Truth of our situation that must have been planned billions of years ago by the master plumber of all energies in the universe.

We have indeed evolved on this Earth. We have evolved so far that our own technological developments are killing us as well as all other life on the planet. We could be completely gone in just a few short years.

But, just as we should all be losing hope, in comes this invader from another realm, the inner realm of our own existence, the Corona Virus and it kills just enough of us to bring us all to attention.

Quickly, we organize ourselves and milatate around the idea that we can outsmart these littlest of things if we just try hard enough and enough of us get involved.

And, very soon after that, we develop the method for controlling these little devils that are still out there infecting us. In a few more years under the current line of research we may be able to prevent this kind of disease and maybe even all diseases forever.

Now, who's smarter - them or us?

My last Zentanglement is about this very important time in our history and you have only to look at this attacker to know how amazing it is to be alive at this point in our evolution.

The Covid Virus is attempting to kill us all by polluting our minds and our bodies. We now have vaccines for the pollution of our physical bodies, but the pollution of our minds is something we have no protection for.

The only thing we can do is come to a united Consciousness. We who know the truth have to inoculate everyone we know who does not have the truth. It's simply a matter of the best of us overpowering the worst of our kind with our superior knowledge.

At this moment in our history there has never been more stupidity and evil spreading across the world. There has also never been so much Truth available to us all. The war is therefore between Truth and evil who's only truth is death of everything. Those who have no Truth are slowly edging us closer to this mass death.

We can and must prevent it. We will do it through the knowledge gained in the last 250 years of Science and scientific discovery such as the new vaccines.

I believe quite strongly that the CoronaVirus has infected the brains of millions of people, but not with the disease itself, with another kind of infection, stupidity, brain decay from lack of use, lack of training, lack of inspiration.

No matter what our political differences, our racial differences, gender differences, ethnic or nationality differences, we must start to come together in a major uniting of our Collective Consciousness Force. Together, we will muster the mightiest force we can bring to bear on our situation in this world.

Perhaps we will all become energized to our highest state by the knowledge that if we don't do this, our time in this universe is nearly at an end. The memory of our civilization will only be known to the people who may or may not discover the plague that was welded onto the Voyager spacecraft, now entering the deepest parts of Interstellar Space.

Instead of knowing us in the flesh and blood versions of humanity, the aliens who run across Voyager and the recording we've stored there, some day in the distant future will learn a few things about our music and our customs. From examining the ship's machinery, they will learn much about our electrical engineering prowess. From the plague, they will have a few important clues about our home planet and the sun that gave us its life-giving energies for billions of years.

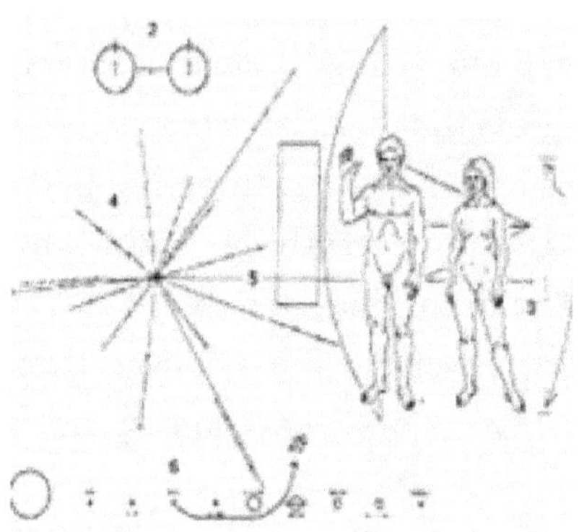

Plaque designed by Carl Sagan and placed on Voyager, with instructions on how to find us, in the long shot case that another form of life may find it.

AND WHAT IF THEY PUT all the clues together and retrace the route that Voyager took in finding them and the aliens are able to make the flight to find us and what if when they found the Earth, there was no evidence that we ever existed?

Is it an acceptable outcome for all the billions of years of our violent and bloody struggle for survival on this planet, only to have none of it remain when the time comes that an alien life discovers our planet and comes to the Earth to pay us a visit but finding only a frozen crust of Carbon Dioxide ice several miles thick covering the archaeological evidence that some partially intelligent creatures lived here?

It's not acceptable to me. How about you? You know where to start?

Start by Zentangling.

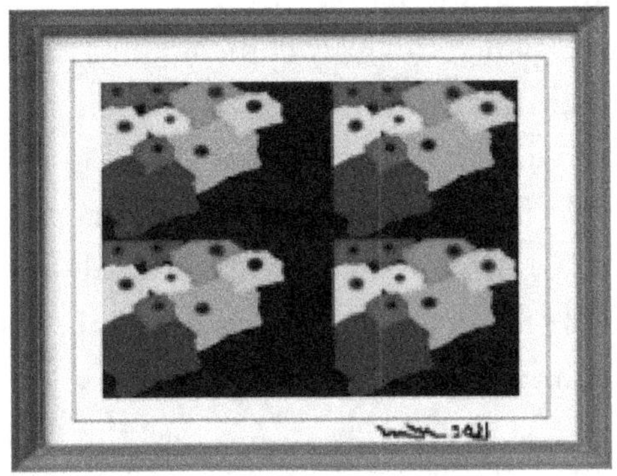

AND ABOVE ALL, IF YOU are ever lost, remember the Three Laws of Consciousness. I guarantee, if you focus for a few minutes on these laws daily, you will help make everything all right. You should even be able to Zentangle by concentrating on the three laws, once again so that you may get them cemented into your own consciousness. I'm writing them again for emphasis and so that you will begin to memorize them so that you can give them to someone else.

Three Laws of Consciousness:

- Consciousness creates all things
- Consciousness connects all things
- Consciousness consists of all things

I hope I don't need to say it – but all of the three laws are deduced from the scientific evidence I've presented so far.

The way that atoms attach to one another to form a substance that we can use like steel, or bricks or even water, is possible because every atom is exactly like the next. There is no difference quantitatively or qualitatively between any of the atoms in the universe or any of their subatomic particles of composition. No matter how many are created to make up the universe all atoms as well as all subatomic particles are exactly the same everywhere. This is how I derived the First Law of Consciousness, the creation of everything at any given moment, always uniformly the same, which can only come out of a major universal Law.

We've already learned about the God Particle in the first chapter. It is said by the physicists of this era that the God Particle Field gives every other particle its mass. And, they use the analogy of particles encountering the God Particle field and suddenly it's as though they were stuck in a field of snow. Now, there is resistance to their movement and this translates into how we perceive mass.

However, it is far more likely to me that the God Particle is not responsible for giving a particle its mass. Instead, it gives every quantum of energy it's size in terms of its energy. Remember that mass and energy are really the same thing.

Electrons, for example, are all measured at: $1.602 \times 10-12$ ergs, or $1.602 \times 10-19$ joules.

Now, this is just scientific notation for making a huge number or a very small number more manageable.

1.602 with an 'X' after it means times, minus 19 decimal places. It's really that simple. So, when it's minus that many decimal places, you know it's very small.

Written out this number is:

.0000000000000000001602 Ergs or about 1 and a half millionth of a millionth of an Erg of energy, or one Electron Volt.

Don't worry, you're not going to get a shock from one electron because this is not even close to the same kind of voltage you get in your everyday electronics which are requiring the movement of billions upon billions of electrons throughout your phone or computer to do the work of your device and these devices of course are designed to use as little electricity as possible in order to be powered by very small batteries.

The only reason I mention this amount of detail in this book is to show that every single electron is exactly - .0000000000000000001602

Ergs. Now, this packet of energy can change depending on where in the atom, or which orbital around the nucleus the electron will inhabit. Every orbital, however, can only happen at set distances which are pre-programmed and never change in the slightest degree in any given atom anywhere in the universe. I classify this exhibit as 'Sameness', part of the fabric of the universe..

This may seem to be a trivial idea, but it's actually monumental because it illustrates how determinative the notion of uniformity can be in order for the universe to work in all directions, in exactly the same manner, at the same time. This denotes that everything that is created comes out of one source, not out of many, not even more than one.

The First Law of Consciousness dictates the uniformity in the universe that we observe as continuity.

Now, if you still don't quite understand what I am saying or remain skeptical, tonight before going to bed go outside your house and look up at the stars and the Milky Way, if you are able to see it in your neighborhood, and realize that what you are looking at are the combination of googolplexes x googolplexes of subatomic particles that make up our universe and not one of them varies one tiny bit from the other one of their same class.

Even the number of electrons and protons that form all of the Hydrogen that is burning in our sun that gives us all of our energy is a number that can't be written down anywhere because there are so many of them.

But, in our Sun, every electron that makes up every Hydrogen Atom are all exactly the same. All of the Hydrogen atoms are also of the same energy as the next Hydrogen atom.

NOW - get this, not only are they the same, they are also spinning in the same direction. They are also spinning at the same speed.

All the atoms in the universe are spinning at the same rate and are therefore synchronized to the same beat.

When a quantity of particles in the universe all beat at the same time they are considered to be synchronized just like a metronome will synchronize with another metronome if you leave them alone.

When this amount of stuff in the universe is synchronized, you have enough evidence to observe the . . .

Second Law of Consciousness.

"Consciousness Connects" The other good news about all of this information is that the 'One-Ness' in the universe is really meant for us to experience. And, again, it's so simple, when everything is already connected to the source, there is nothing more to do for the creator of all things because all things were created connected by default. This kind of creation on all levels and all dimensions is of course, far beyond any kind of creation that we humans can even dream of.

Any one of us has the ability to experience it. It matters little whether you are young or old, male or female, short or tall, smart or dumb, any one of us can experience the 'One-Ness' by merely blocking out all other thought and waiting for it to come into our minds. You'll know when you've reached it - trust me.

There is no other way to describe the One-Ness for this book because every individual may experience it differently. But, it's easy to understand, now that you are familiar with the total flow of energy in the environment we all share.

The Third Law of Consciousness - that Consciousness 'Consists' of all things – is a bit more gritty, yet still simple for anyone who is following the bread crumb trail I'm leaving.

And, when you feel the Creative Force that puts all of these things together in such a nice neat little package, you are also ONE with the Creative Force and almost anything you desire to accomplish in your life is now within reach. This is only possible because Consciousness is alive and well inside of all things, but especially in all living things. Since you are a living thing, Consciousness consists inside of you. How much of it is your own prerogative and by that I mean, you can turn up lots of it like stepping on the gas pedal or you can put on the brakes, stifling Consciousness to get through your day more easily. We can always choose how much or how little Consciousness to open up and rely on in our daily lives. This is what we feel as 'Free Will'.

The Creative Force is no different when we are the creators of things or when the Creator of the universe is doing the creating. The Force is the same. The only difference between our own little amount of The Force that we use to re-create in our lives and the amount of The Force that the universe requires to remain open to us is in the amount of the electro-motive force, the light force, the gravity, the Strong Force and the Weak Force that holds all of our atoms together and the amount that the creator of the universe has available to him, her or it, which is of course an infinite amount of power. For humans, the most electro-motive power we can generate is barely enough to light up a small light bulb.

As human beings, we obviously cannot create as much as the master Creator of all things. But, we can experience the same pleasure and the sense of achievement that he, or she, or it, the Creator experiences when creating the universe. Just not in the same amount. This is another reason why we are chosen to be the 'Feedback Loop', the fourth and highest 'State' of Consciousness.

And, because of the Second Law of Connecting, we know that all of this is true or can be true if we simply put a few brain cells to working on it. As night follows day – it will happen.

We can be the same as God, just not the equal of God. When I create my books, I experience it. I bet that most other authors, when creating their books, feel the force of Creativity and are deeply moved by it.

When Paul McCartney was interviewed one time about how he created so many songs that have become so universally beloved by millions of people, he said, "I don't know where it comes from. It's just like it comes down from out of nowhere and goes through me."

I have heard similar things from other song-writers, too many to mention. Artists feel the same way when they've painted a beautiful painting. When I fix the house in some creative way, even the plumbing, I feel the creative force within me and it's one of the most satisfying and long-lived feelings I've had.

Little did I know that when I sat down to doodle around with Zentanglements that I would find my most creative moment in my life, but I did.

In short, to keep it simple. If you really want to experience life, just keep on consisting in it.

Chapter Four

- The 4 States Of Consciousness -

AT FIRST BLUSH, IT may seem contradictory to you right now that there would be 3 Laws of Consciousness but also 4 states of Consciousness. This is easily resolved when I tell you that the 4 states of Consciousness apply to all things that exist and have been given their role in the universe. And, this implies that all things that exist and have been given a role have been given that role by another, higher force. This must also be true, because we simply cannot conceive of a universe where everything, all the google-plexes of things multiplied by the googolplexes of things all getting this information on their own.

Therefore, the 3 Laws Of Consciousness apply to the originator of Consciousness and that would have to be God or something like a God. As part of God's creation, we can only admire the job God has done through our observations and calculations. This is the job of Science and as such is part of the highest form of Consciousness, in this realm, the only one we can be sure of as you will soon learn.

You have learned about the 3 Laws of Consciousness. Now, you will see how these laws constructed the entire universe out of whole cloth, or actually out of the 4 States of Consciousness, a reality almost automatically induced by the 3 Laws.

The most common debate in Physics used to be whether or not there are four fundamental forces or three and might we someday be able to get them all down to one. The four that we're used to hearing about are:

The Strong Force

The Weak Force

The Electromagnetic Force

Gravity

The Strong Force is the force that holds the Protons together inside the nucleus of the Atom. This is the strongest force we know because it has to counter the tendency for two similar charges, the positively charged protons to repel away from each other. If they did that by overcoming the Strong Force, there would be no atomic structure anywhere and the universe would simply not exist.

The Weak Force is the force that is responsible for all atoms to decay.

The Electro-magnetic force is the radiation that we can all feel, see, taste, touch etc. It is light, electricity, radio waves, television, Internet bandwidth.

Gravity, we all know about from the earliest age when we fall down or bump into something. It's the force of gravity that we feel everyday. Gravity is actually the weakest of all the known forces, but it sustains itself over the largest distances. We all know how the Earth is in a stable orbit around the Sun because of the Sun's gravity holding us here. And, that's a good thing. But it's also the force that holds the Sun in orbit around the center of the Milky Way Galaxy. The Milky Way Galaxy is held in place where it is and where it is orbiting by the same force of gravity that holds all galaxies together in the Cosmic map that resembles a huge brain.

Gravity is the weakest of the forces but affects everything in the universe over the largest distances we can measure and the Strong Force is the strongest of the known forces but exerts itself only as wide as the diameter of the nucleus of the atom because it's only in this smallest amount of space that it is needed.

But if this strongest force in the universe went any further, or had any influence any further than the diameter of the nucleus of the atom, this would be a very strange universe where we were all stuck to the atoms by the Strong Force instead of stuck to the planet by the weakest force of Gravity.

Indeed, life might not even have had a chance given this extreme size limitation.

But, I digress.

More recently, the most common debate is whether or not there is a fifth force.

Everything we know is all about the 5% of the universe that we can see out there and the goings on inside it all.

The other 95% of the universe, what we are calling the 'Dark Matter' or 'Dark Energy' we know absolutely nothing about. Worse, it appears that this 95% of the universe that is dark to us is pushing the universe out into expansion that is accelerating faster and faster every day.

We know this by simply pointing our telescopes at the furthest objects we can observe and record their speed away from us. Ever since we started watching the edges of the universe like this with Edwin Hubble, we could see that the universe is expanding and accelerating the expansion in every direction.

This is happening when all our knowledge of Physics and the laws of gravity are telling us that the universe should be slowing down and collapsing back in on itself because of the law of attraction of the massive bodies, planets, stars, asteroids, galaxies, black holes, etc. that have been compacted into Space since the beginning, each of them having a pull on one another. The more of them that are created, the greater the force should be for the universe to collapse or at least slow down the expansion. But, in fact, just the opposite is happening.

Black Holes the darkest of all matter that is known to gobble up entire suns and sometimes even whole galaxies and which serve as the center of all galaxies we can see out there, are also thought to contain all the information that the universe needs to reconstruct itself should the need arise. Are Black Holes some kind of Backup Device for the universe's memory? If so, how is the information stored? Could it be that the end result of immense gravity that is contained inside a Black Hole is a form of Consciousness strictly used as a zero point of no return for the universe as reference points?

Everything that we call Dark Matter seems to be connecting all of the galaxies by this weird stuff that we know nothing about and more than that appears like filaments that connect all of the galaxies together in some way. And shown in a larger perspective, this strongly resembles the structure of our brains.

Now, let's put two and two together, because no one else is.

A. We have something known as Entanglements where the smallest of particles are connected to each other in some unseen way, could be via unseen strings.

B. We have an unseen force that connects all of the galaxies together in a string or filament type of formation.

Putting these two observations together, what pops up in your mind?

I know, right?

Let me clarify.

From the smallest things to the largest things in the universe, there appears to be a connective tissue of some kind. We know absolutely nothing about each of them. However, we can look at the world and see that there are connections that are working in mysterious ways.

When we look outwards with telescopes, we see something unexplained, the 95% of the universe that we don't know anything about doing the opposite of what we would expect.

When we look inwards with microscopes we see something also unexplained, perhaps as much as 95% of the workings of the Quanta, the smallest things we can observe, that we know nothing about and also doing the opposite of what we would expect.

Next, put it all together, things we know nothing about and doing the opposite of what the laws of Physics require, are probably all part of one thing, one force, one of something that is everywhere.

It's logical, is it not? To believe that the two connective tissues that we've discovered at roughly the same time in our development, may themselves be connected to one another?

Is this the Fifth Force? If it is the Fifth Force is it dominant over all the others?

If we observe it happening in 95% of Outer Space and probably 95% of Inner Space, isn't it logical to assume that it is dominant?

I promise, by the end of this book, you will have the answer.

#

WHEN I FIRST LEARNED about Entangled electrons and other subatomic particles, I had to ask the question: 'How do they know?' How do they know about the existence of their partner? And how did they know what to do when their partner did something else? And, is this a form of 'Electronic Consciousness?'

It has now been shown that some species of Bacteria, probably the earliest form of life on this planet going back about 3 billion years when life first began here on the Earth, have the unique ability to digest electrons. Life, at its earliest form is so close to the electrons that they can eat and digest their energy in order to stay alive.

They are called Shewanella Oneidensis - shown here.

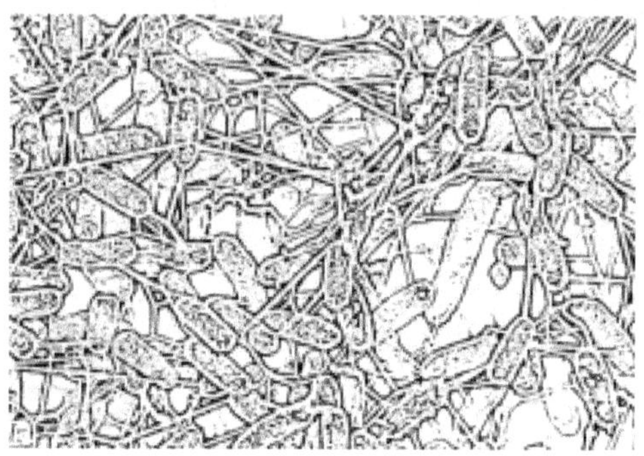

AND Geobacter shown here.

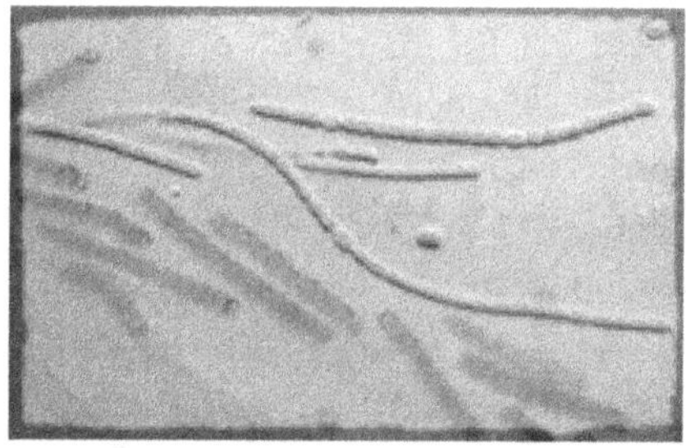

AND HERE IS A SINGLE garden variety of bacteria, billions of them that look exactly like this inhabit your gut and do the work of digesting your food and turning it all into chemical and electrical energy that powers your entire body.

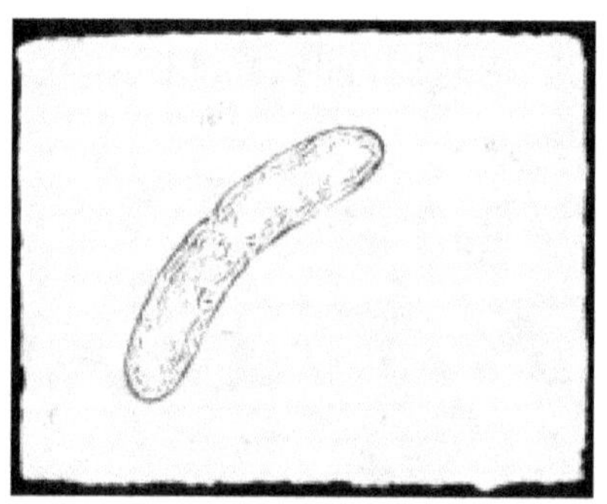

A single bacteria less than one micrometer in length.

WHEN YOU UNDERSTAND that more than HALF of our human DNA is exactly the same as Bacterial and Virus DNA, and that they can eat and digest ELECTRONS, I take that as very strong evidence that our own DNA, descended from Bacterial DNA is the basis of our own State of Consciousness that we have evolved with over the last several million years and that we owe to these creatures.

We humans do not eat and digest electrons directly, but after the final process of digestion is done deep down in our guts, electronic energy is what flows up to our brains and keeps it powered up. We are living computers made out of flesh and bones and supporting organs, but it is the power of electricity that moves all of our parts at our beckoning and it is the beckoning of the brain through the movement of electrons all through it that produces the thoughts that make us move around the planet as we do in so many fantastic ways.

As you will learn or you may already know that your consciousness, the thought patterns that make you who you are, or at least who you think you are, didn't just arrive on the scene the day you were born. Actually there were billions of years of Evolution that had to take place and that goes into the chemical mix of what you are, what you do, how you think and even the major events of your life. We've already seen clear evidence of this in something called the 'Zinc Spark', in an earlier chapter.

None of us happened over night. And as we have already seen from the lessons in the last chapter, the major structure of ourselves, our DNA, has been around for billions of years. Your DNA, the code for how most of your life's events will transpire isn't created on the day you are conceived by your parents.

The day of your conception is a very crowded mash up of coding that came to you from your two parents, their four parents, your grandparents, their eight parents, your great-grandparents, your sixteen great, great parents and so on and so forth where eight sets of DNA coding is pre-manufactured by the sixteen parents who came before the eight grandparents. The sixteen predecessors to you, manufactured from the thirty-two who created them, makes the numbers of your DNA contributors go up by twos every generation, until it reaches into the millions of contributors of what makes you.

This is just like when computers were initially invented with 8 kilobytes, and then 16 kilobytes which then advanced into 32 kilobytes, or 32 thousand bytes of information, soon replaced by 64 Kilobytes, and then, 128Kb machines, 256kb machines etc. Then, we quickly manufactured machines based on Megabytes or millions of bytes of information. Today, we operate computers of many Gigabytes, billions of bytes and Terabytes, trillions of bytes and so on, until we were recently forced to invent a larger number – the 'Flops and Gigaflops' to contain all of the jumps in capability. But the jump in memory for the computers we use today is an excellent analog of how all life evolved on this planet. Nor is it an accident that we would design machines that mimic our own evolution in this way.

You can see that there are millions and even billions of humans who make up your DNA. They are all a part of you, like it or not. None of us can escape this destiny. It's how we are created because it is how life was created billions of years ago.

Bacteria, created as the first form of life on Earth about four billion years ago, are like the first personal computers, the TRS-80 or the Commodore 64. They would contain just enough DNA, just enough code to perform the simplest of operations.

From this humble form of life that probably took all of its energy from electrons that flowed all around them from lightning perhaps or other natural electrical events, more complex creatures evolved who would not be able to ingest electronic energy directly and would be forced to grow digestive systems that had other food items, simple proteins that would in turn, give up their energy easily inside the guts of these early animals.

From these we evolved over billions of years to become massive towering creatures that wouldn't even have the ability to see our forebears without the use of scientific instruments. We wouldn't even know these predecessors existed at all without our advanced cranial abilities that slowly became more and more powerful, inventing more and more powerful machines and mechanisms.

There's an old adage in Science that has since been discredited and almost forgotten.

"Ontogeny recapitulates Phylogeny". It refers to the notion that babies in the first few weeks appear to be reptilian and so it seemed to scientists of a century ago, that perhaps our entire evolution was re-enacted in the womb as we became more human over the nine months of our gestation and up to the time of birth.

But, perhaps a better use of the phrase is that our own Evolution of creation, the machines we invent, does resemble in many ways and mimics the evolution of ourselves.

We've even arrived at the creation of something known as 'Artificial Intelligence.'

Many of us are now asking a very cogent question: "Where will 'Artificial Intelligence' lead us?

THE ELECTRONIC STATE of Consciousness

Before going any further, and after talking about in the three previous chapters, I think it's a good time to define exactly what Consciousness is, or what a 'State of Consciousness' could be.

And, the simple way to grok this is to ask yourself the question that most of us ask at some point in our lives.

"Who am I, really? Am I more than what it says on my drivers' license or birth certificate? Am I more than the accumulation of memories that I've made over the years that is now a fundamental part of my 'Reality'?

AND, "What is the source of the voice in the back of my head that I am forced to listen to every day and night of my life thus far?"

OR, "Who am I, really?"

And, the answer to that question from the perspective of this book is that if you are a living human being, you begin life by asking these types of questions, by default, even though they may be phrased as 'What's for dinner?' or 'Can you find my shoes?'.

As soon as you realize there is something going on that is external to you, you have accepted as reality the artificial barrier between your own little manipulation of the universe and the rest of it. This is Consciousness.

From that moment on, the emotional and mental distance between your own existence and the existence of the rest of the universe widens. You have become an individual. It's the recognition of this fact that is the basis of Consciousness.

Life happens and soon, the differences between you the individual and the rest of the world become more numerous. By the time you are a young adult, you have more than enough evidence to recognize that you are someone special and you expect to be treated as such by the rest of society.

The way that we battle these elements of the rest of the world arrayed against us and who also want to be treated special and gain an advantage over others like us, is what makes up our personality.

Personality, however, has nothing to do with Consciousness, nor does Intelligence have anything to do with Consciousness. These are merely capabilities from our brains produced from our DNA.

But, our Consciousness begins like a tiny seed planted deep inside our brains from the moment of our conception and grows slowly over time to become aware, not only of the external stimuli like food, parents, teachers, ministers, law enforcement, political leaders, friends and all the myriad forms of entertainment, but eventually begins to look much further in the environment for deeper answers.

How far we allow our imaginations to roam out there determines our depth of Consciousness, our measure of our Consciousness, which someday may be measured like we do our Intelligence Quotient or IQ.

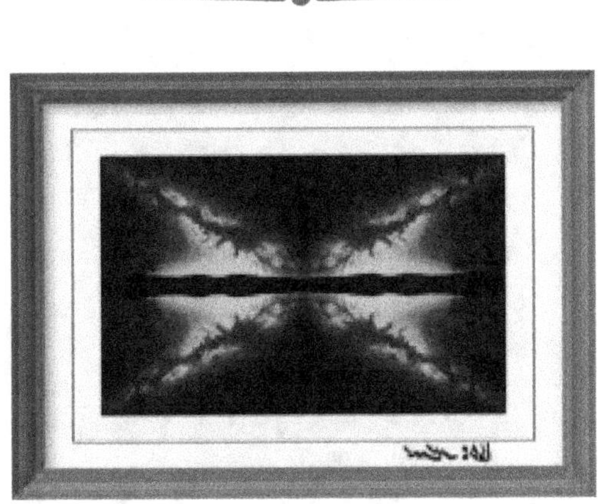

Looking out and wondering about the Milky Way

SO, TO SEE HOW LARGE your Consciousness can be, we're going to remember the Electronic State, where it all begins.

And so it seems that there are four and only four states of Consciousness and the state of Consciousness that you are using right now to read these words is part of the 4th and highest state - what I call the Feedback Loop. We are here because a true and deep form of Consciousness can not exist without feedback from something external to it. Consciousness must ask questions and it must receive some answers in order to be fully aware of itself in consideration of all the answers. Thus, all of life in this universe evolves as the necessary outcome of the questions and Consciousness finds a way.

And, I hope we don't have to debate about the idea that dogs and cats, and even pigs and goats have a Consciousness. All you have to do is to sponsor one of these wonderful animals to know their love and devotion.

Dogs are much smarter than most of us know.

We've all heard, or read about, or seen in the news, amazing stories of dogs saving the lives of their owners, or giving their lives to protect human children. What possesses these wonderful creatures to do this?
So many dogs have given their lives in service to their human masters as K-9 units in War or in service to the blind or handicapped.

IF YOU HAVE EVER BEEN a cat owner, you also know how intelligent and aware that a cat can be.

You are struck with the contentedness and relaxed nature of this animal and the funny and lovable things they do.

Satellite view of my house in Santa Cruz

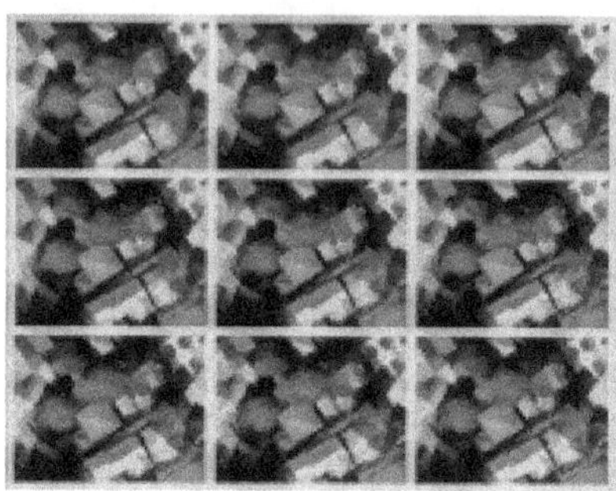

My House by Satellite - Zentangled
Why is my house viewed from the satellite a part of this book? I have
no idea, except that I'm always doing my most creative work from
inside this wonderful house. It could be tapping me on the shoulder
for a little recognition, perhaps?

———— ◉ ————

And if you have watched bees buzzing around in your garden or
studied them in any manner, you know that even insects are aware of
something far beyond their own little lives. As we all know, bees are
extremely hard little workers for the benefit of the hive. They
unselfishly live to do their duty in support of the next generation of
bees and then expire.
Without the bees, by the way, there would be no pollination and
therefore no reproduction of the plants of the planet that we require
as food.

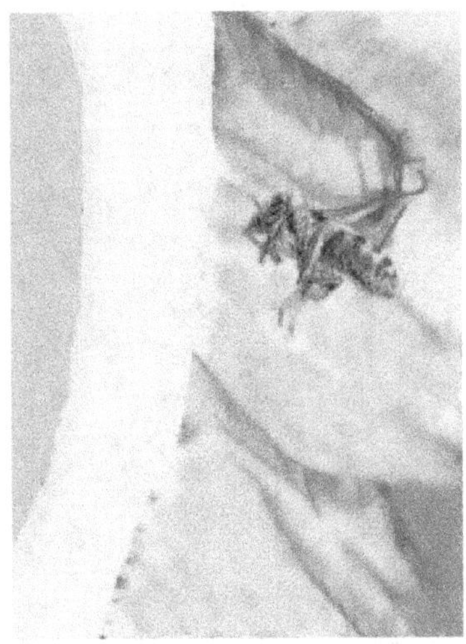

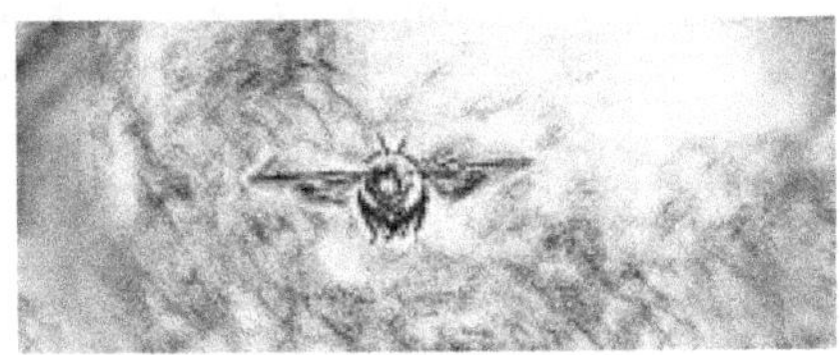

The bees' appreciation of geometry and construction in building their honeycombs is beyond what most humans possess.

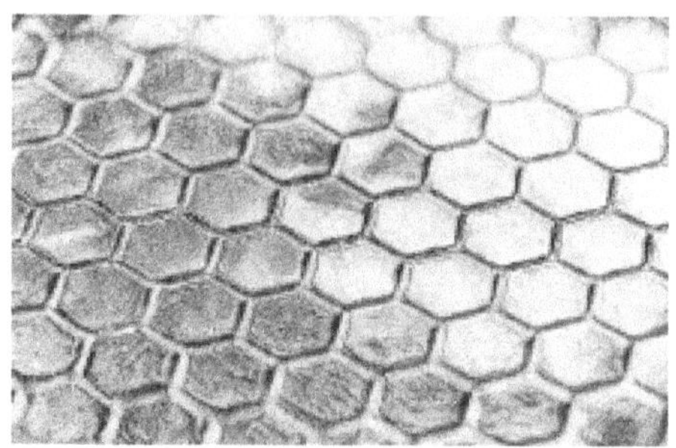

So, most of us would find it easy to believe that most insects possess a kind of Consciousness.

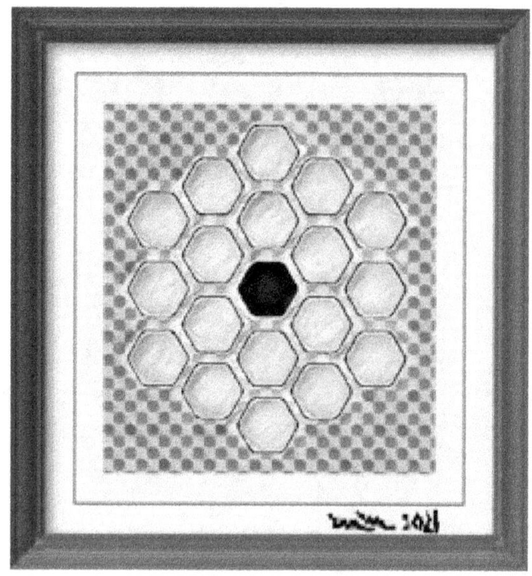

The Webb Space Telescope

ALL LIFE
MATTERS

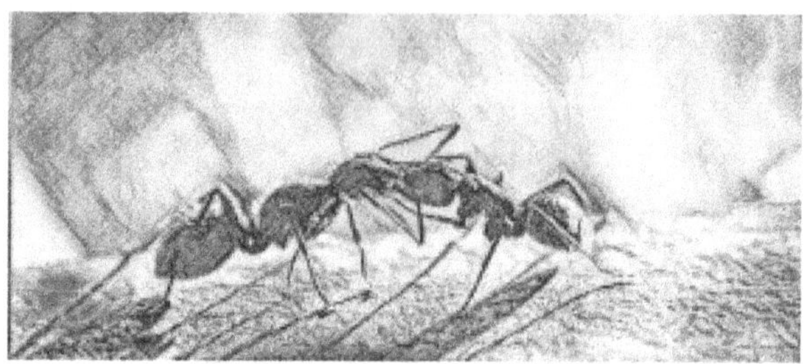

IF YOU'VE EVER WATCHED ants as they go about their daily business on your patio or in your backyard, or out in the jungle, you've noticed that every time they pass another member of their colony, they must greet each other and rub their antennae against the other's. It's obviously a way of recognizing one another and giving their 'How you doin?'

Most ant species have some kind of shared hive where they congregate, but some species go even further, enslaving other ant species and forcing them to farm the fungus that they grow on their 'Plantations' – the earliest form of slavery?

AND, OF COURSE, MOST of us find the activity of birds singing to each other, getting together and mating in the Spring, diligently making nests, feeding their offspring as definite signs of Consciousness.

I have several bird houses and feeders on my property. I feel so blessed to be able to do that for them. All over the world, birds and bees are under intense deadly pressure, caused by humans, that is making their numbers decrease by the millions each year. As the world becomes more silent, without their songs to wake us up in the morning, is it getting any better?

But how about trees and other plants of the Earth, most of whom have been around much longer than ourselves.
HAVE YOU EVER WANDERED through a Redwood forest?

If you have, then you know the peace and tranquility bestowed upon us from above that only a Redwood forest can provide.
New research has informed us that trees actually communicate with one another over a network of trails made by fungi, mushrooms living in and among their roots.

They never brag about their age, but some of these venerable creatures
have been around since before Christ.

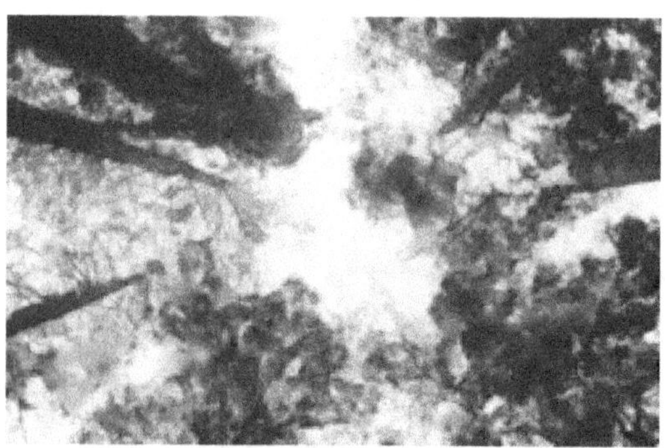

A few years ago, I had a few minutes to spare and I wandered into a
Redwood grove of what I thought to be a family of trees. I looked up
and the wind blew them around at the tops far above me. They seemed
to be dancing to the silent music of the wind and singing to each
other. It was a memorable encounter.

PERHAPS BECAUSE OUR own self-centered way of looking at the
world keeps us blinded from the truth that other species of animals all
the way down to bacteria probably have a level of Consciousness, we

think Consciousness is something unique to us as a species. Some of us even believing that it is exclusive to us as a person.

If electrons and protons and photons and quarks have demonstrated properties of knowing about themselves, then, it's quite a simple leap of faith to realize that all life forms on this planet and probably on all other planets have the same quality of knowing just enough about themselves to carry on in the manner for which they were designed.

What may make us different from all other life forms is that what we were designed for is vastly more significant for the rest of the universe than the activities of any of our partner species who exist on the Earth mainly to be our food source.

It could be we're missing the most important ways that other animals, insects and plants can nourish us.

WE MUST NEVER OVER-think Consciousness. It's that simple. We should all obey this rule because it is taught to us by all other life forms that we know. Cows are not thinking about their lives every minute the way we do. They simply chew the grass and digest it so that we can have their milk and eat their meat.

The same thing with chickens. If animals that we use as food had too much awareness of their situation, they would all be so anxious, they would never get fat and gain weight and everyone on the planet would starve. They don't have this kind of anxiety because they have no idea what is going to happen to them in a matter of days, weeks, or at the longest, months. If they do, they don't seem to care.

The Rotational State of Consciousness

When you look out at the universe and stare at it with the most powerful telescopes in human history, you find that everything is rotating or spinning. There is nothing that we see out there that is not in a rotation about an axis.

Everything we look at under a microscope is also rotating on an axis or spinning. There is no 'thing' that we can observe in this universe that is not in some kind of rotation around something, as the Moon rotating around the Earth, the Earth rotating around the Sun, the Sun also rotating around the center of the galaxy.

Electrons rotate around the nucleus of the atom. The photons are spinning around on their axis and even the Quarks that compose the Protons are spinning and spinning more rapidly. If there are any particles that compose the quarks, they will be found to be rotating or spinning in place.

Everything, then, in the universe experiences what I call the 2nd State of Consciousness or 'Rotational State of Consciousness'.

Why do I call this Consciousness? Because this 2nd State of Consciousness must be inherited from the 1st State of Consciousness by what Scientists call 'Induction'. Just as the electrons are aware of each other, this awareness or the ability to be aware is passed up to the atoms that the electrons will form. The Rotational State of Consciousness is induced by the 1st State, in other words, given to the next level of existence, the atoms, because this is the nature of angular momentum – one good spin creates another.

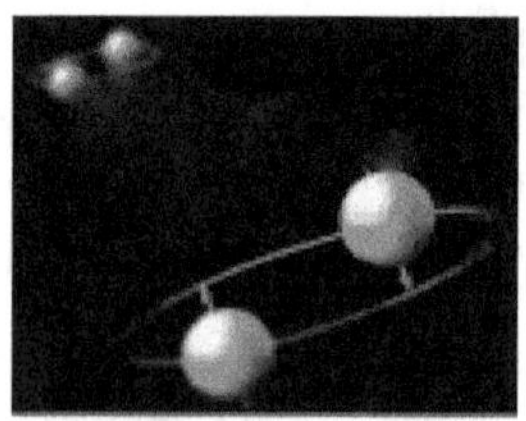

When Entangled - one electron Spins UP and the other spins DOWN. When either of them changes direction, the partner also changes. How do they know about the other?

MY ZENTANGLED VERSION of Entangled Electrons in a collage.

If you don't believe in this magical concept called induction, then just look at your newer phone chargers where you simply place your phone on top of an 'Induction' plate and while your phone and the charger plate are in close proximity to one another your phone's batteries are slowly recharged.

How can this happen without recharger cables? Through simple induction. Electrical force produces magnetic force that in turn produces electrical force and so the electrical force of your recharger plugged into the wall flows into your phone's batteries without the old-fashioned cables.

The way that light travels through Space at the speed of light is another common everyday example, because we all use light throughout our waking state every day of our lives.

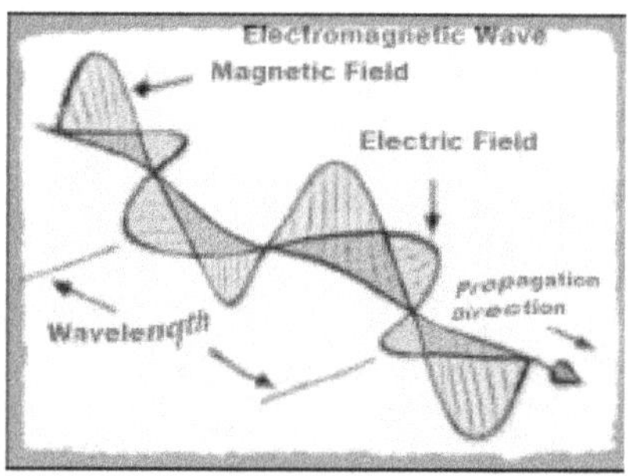

Light waves travel all over the universe through the process of induction of one form of wave energy into another.

THE VIBRATIONAL STATE of Consciousness

We all know that atoms clump together to make molecules, which must consist of at least two atoms. Molecules also do not remain stationary in space but are in constant state of motion. But, the molecule doesn't rotate as much as it vibrates. It contains the energy of the atom but stored in a spring that can be released as energy at any time.

Once again, this is produced by the process known as Induction. The atom gains its rotational state from the electronic state of its electrons and the molecule gains its vibrational state from the rotational state of its atoms. Since we are all made out of molecules, we living things gain our state of Consciousness from the Vibrational state held captive by the molecules that make us human. Music, by the way is another form of vibration we all live for.

Everything in the universe that has Consciousness seeks the next highest level throughout its life. This basic, universal desire is what makes the universe into the wondrous and beautiful thing that it is.

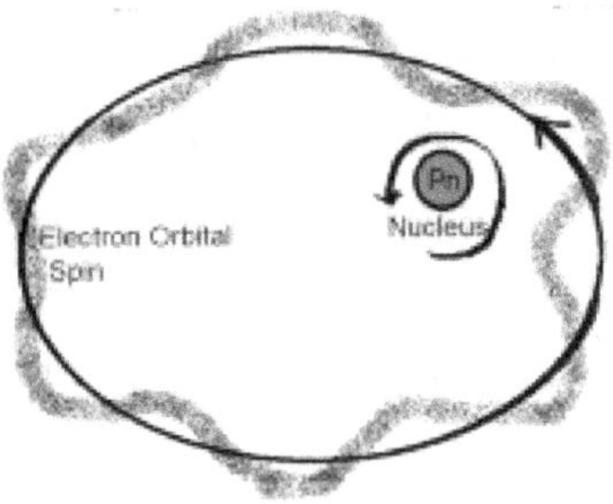

Do the Protons in the nucleus KNOW where their electrons live?
TRY THIS NEXT TIME you have a chance to meditate. Try to imagine the tiniest electronic device you have that possesses even modest amounts of Artificial Intelligence like Siri or Alexa. Then, take this level of Consciousness and apply it to any living thing. What are the differences? What are the similarities?

And now is a good time to talk about Synchronicity or the synchronization of all things.

As we have learned, all electrons in the universe have exactly the same size or wavelength. All protons in the universe take up exactly the same amount of space as the other ones. All neutrons, all quarks, all photons, gluons and muons, every fundamental particle in the universe is the same as any other of the same variety. This is due to the synchronization powers of the Creative Consciousness of the universe.

This process has developed over billions of years since the Big Bang, and therefore, not everything in the universe is totally synchronized just yet. For now, you must realize that everything is attempting to be synchronized. This is why induction works so well. What is synchronized in the lower states of Consciousness wants to be synchronized in its next highest state and so synchronicity is maintained every step of the way, all the way up to us.

Ask yourself when did you experience the best times in your life and the answer should reflect a time of great synchronicity, when everything just seemed to be flowing according to the way you would want it to be flowing in your most insightful moments.

Synchronicity is how I was able to learn these things. From your own encounters with it, you will learn other things and over time we will all share in this new knowledge and our individual lives and our society will evolve.

The 4th state of Consciousness – The Feedback Loop

Have you ever tried to chant 'OM' to yourself in the deepest monotone your vocal chords will allow. If not, this would be a good time to start.

Take a deep breath and continue the tone like in an 'OM' expression and control your breath so that the tone continues as long as a minute or whatever period of time you can do. Practice this until you can double your time of holding the tone.

When, you believe you're at your maximum length of the vibrational tone, focus on the molecules inside your own body that are allowing you to perform this miracle of life, carrying air from the lungs, which is produced by the molecular structure of the lung tissue and then flowing over your vocal chords in such a way as to become one with your molecular Vibrational State of Consciousness.

This is harmonizing with the universal states of Consciousness that is the basic fundamental reason why we exist. If you can feel what I'm talking about here, you're going to get very relaxed and content, happy even, and without drugs of any kind.

Which brings me to the 4th and highest state of Consciousness as far as any living creature can attain and that is the Feedback Loop.

What you are doing right now, as you read these words, as you consider the concepts held within it and other materials of learning is giving feedback to the universe. Whatever your reaction to all of the wonder in the universe, the universe perceives it as 'Feedback', exactly like your ratings for the things you buy on Amazon.

And this is an automatic feature of your sensorial organs. You can't help but give feedback to everything that you see and experience in life and as such your feedback is honest and candid and solid and completely and individually your own.

Feedback by living things is the reason that living things exist. The amazing and miraculous things in the universe that we witness are things that the Creative Force of the ultimate Consciousness cannot see, feel, taste, touch or hear because there are no sensorial organs attached to Consciousness.

The sights and sounds and smell and touching of things in this universe are experiences that only creatures that are alive can know because we have been gifted with our sense organs. Remember, nothing comes out of nowhere. When we experience these things, the sensations are transmitted back to the Creator because all of these energies are a two-way street and the Creative Force can't really experience all of his magnificent Creation the way that we can see, smell, hear, taste and touch them.

No one is immune to it, not even the lower animals. Everything with a brain resonates at some harmonic scale with the lower states of Consciousness, and so, even if we are not paying attention to the Creator, he, she or it, gets the information it needs to improve our lives, our destiny, our evolution.

The more you appreciate this role, this purpose for your life and the more you find yourself in a contemplative state, the more your information is gathered up and transmitted to the Creative Consciousness, the connecting tissue in, amongst, and in between all things. Remember Einstein's 'Fabric' of Space/Time?

This is who you really are. You are simply part of the Feedback Loop, the way that your Creator takes back recognition of his, her or its gift of life. We are so designed that we cannot avoid giving the recognition of life back to our Creator who cannot learn of the things that happen to us without our feedback.

This is why we are not designed as mechanical machines or robots, something the Creator could have done. Instead, he created us out of the energies that swirl about in the universe via several levels of Consciousness, the very essence of his, her, or its existence.

Don't believe me? Try it just once and wait a few minutes, or a few hours and see what happens as a result. Try being the electron as it spins about the nucleus of the atom. It's not that hard. It's actually the easiest thing you'll ever accomplish – trust me. Then, try to feel the same atom friends clumping together to form the machinery that makes you tick. You'll hear your stomach growl. You'll hear the noises about your house or your neighborhood, all part of the music of life.

You only have to do this for a few seconds or at the most a few minutes. Listen to what your mind tells you about these sensations and try to decipher who is listening to this, besides your own self. Can you feel your heart pumping? Can you hear your blood flowing through your veins? Can you watch your immune system battling against all of the invading viruses and bacteria of this tiny world of competing armies?

Next, wait for night time. Go out in your backyard with a pair of binoculars or a telescope and look up into the night sky.

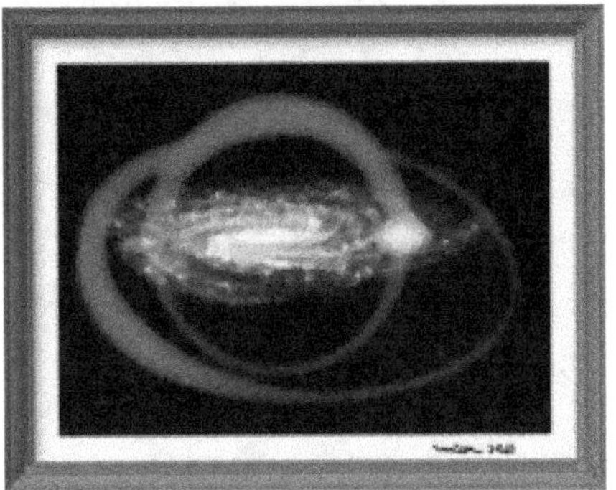

Our Milky Way Galaxy with new revelations from The Gaia Space Telescope

Wait for the vastness of Space and Time fill your brain, your heart and soul and give thanks to whatever it is that keeps all of this in place for you.

REALIZE THAT YOU ARE the reason for all of this to be happening. It is not, never has been and never will be an accident.

And now one more time just to review. It's so simple, anyone can place these thoughts in their memory and Zen meditate on them. Try it for just a few minutes or better yet, do it with your own Zentangle. You'll be glad you did.

Three Laws of Consciousness:

- Consciousness creates all things
- Consciousness connects all things
- Consciousness consists all things

CONSCIOUSNESS CREATES, connects and consists from within everything that we can see in the universe and probably in everything we can't see – yet. This is the greatest Zentanglement I can give you. Your challenge is to figure this one out on your own terms, with your own unique skills.

Until the publication of this book, no one really understood how it is that every single electron in the universe contains exactly the same voltage. Or how is it that every proton contains the exact same electrical charge, just enough to keep the electron in the exact place it needs to be, to within micro-micro-micro millimeters.

The Three Laws of Consciousness keeps them all the same size and energy to the 'nth' degree of accuracy, because it can. And so we can gather from this evidence that Consciousness prefers the universe all nice and neatly organized in this way.

The other final thought I have for you is that the Covid-19 Virus has taught us a great deal in the last few years. As I finalize the words to this book, the Covid-19 Virus has mutated once again and made itself more virulent and contagious, possibly even more deadly. This is after the most recent human reaction has been to create amazingly effective vaccines using the virus's own DNA to train our immune systems how to fight it.

But, when the virus learns of this latest technological leap on the human side of the battle, the virus is able to outwit us again and it's back to the drawing board for our drug companies to make yet another technological breakthrough. It's almost as if the virus life form has figured out that its mission is to eliminate as many humans as possible because there are too many of us living on the planet today as each one of us killing the planet just a little bit more each day of our lives due to our heavy carbon footprint. We can't help ourselves. The politics of our civilization dictates that no one can really do much about this mass suicide and so the virus level of life on Earth has taken it on itself to make the necessary adjustments. Viruses don't play politics. They have not evolved to our level yet, but they do play a role in controlling the numbers of higher life forms who have yet to learn how to control themselves.

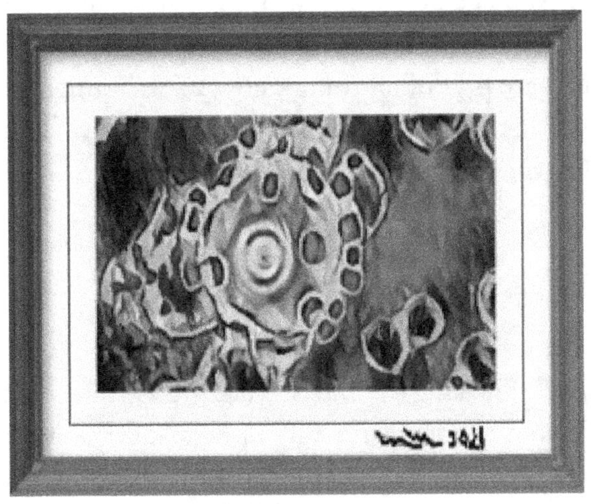

The Covid-19 Virus – Zentangled with an Atomic Blaster

NOW THAT YOU HAVE FINISHED examining my work, hopefully absorbing most of it - when you look around at the universe that engulfs and consumes you every minute of every day, please remember one word before all the others that rattle around in your brain.

That word? 'Zentanglements'. In other words – How everything is Zentangled. Imagine all, or as many as you can, of the most energetic and uniform artifacts of God's hand constantly in motion, creating all the reality that you see just as we create our computer world out of tiny pixels.

What making Zentangled artwork has taught me is that you can see the universe in a coarsely defined and muddled presentation if that's as deep as you want to go. Go a little deeper, however, and the pixels that make up the scenery presented to all your senses will become deeper and richer and flooded with more and more pixels with greater and greater resolution, deeper perceptions.

From this vantage point, you will start to see how Consciousness creates all things, how it connects all things and how you and everything else are part of all things, how Consciousness must consist of all things, including all of the little pixels of awareness and Consciousness that glows inside of you.

———————◉———————

ALBERT EINSTEIN ONCE said that "The most incomprehensible thing about the universe is that it is comprehensible."

It was utterly amazing to Einstein that the workings of the universe are so simple that anyone can really understand the most important aspects of it. His major contribution was in showing us how simple space/time really is. It's just a fabric – who knew? It was his great genius in that he was able to prove some of these strange things to the rest of us. But Einstein died knowing less than you do now.

———————◉———————

IF YOU FORGET EVERYTHING else I've taught you, remember, at least this one thing – In the beginning - you were celebrated in the Spark of Life - The very FIRST encounter with this form of the first state of Consciousness, but far from the last one.

———————◉———————

Now that you are armed with all of this information, and the three laws of Consciousness, the worst thing that can happen to you is that you will find yourself more focused than ever before in your life. YOU WILL FIND THAT you can perceive reality in finer and finer detail. It's like finding a new pair of eyes that can detect things that you see with greater and greater resolution. The normal world of 3 Dimensions may even evolve to that of the 4th Dimension when you can experience Time actually stopping.

If and when any of this happens, your wave energy may be traveling at the speed of light.

AND SO IT IS – that you now know more about the universe than Albert Einstein did when he died in 1957. He was not aware of the Zinc Spark of Life, nor was he aware of the 3 Laws of Consciousness.

And if you've been paying attention, you know a whole lot more than that.

The Lightning Bolt Nebula

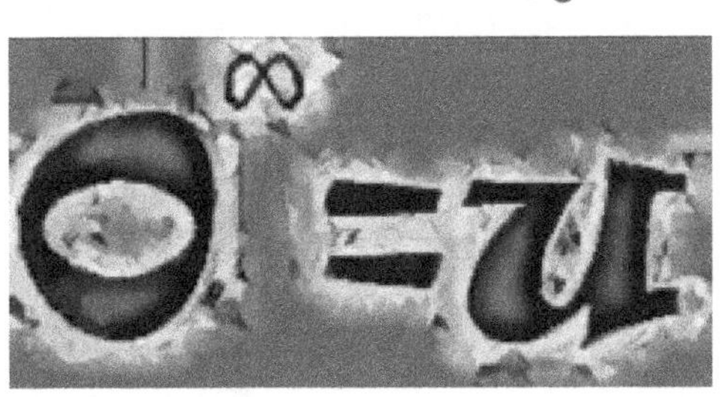

The Proof

UNTIL NOW, NO ONE KNEW how it all snaps together. It's the one-ness of all things that keeps it all together. You have the mathematical proof rattling around inside your brain now if you took the time to understand my new formula.

Extra points if you can explain it to a friend.

AND FINALLY - my Zentangled Illustration of how Consciousness precedes The Big Bang.

On a Personal Note:

I AM ADDING A SHORT note to my readers that did not appear in the first edition of this book. This has come to me during the last days of the year and into the first few days of the New Year – 2022 while I was working on perfecting some of the illustrations and becoming even more intimately involved with them.

I'm in my mid-seventies now. And, every day seems to be going faster and faster in my life. I'm hoping you've heard about this phenomenon from your older relatives. Sometimes, they will express it differently by saying that 'Life is short' or that it seems that they are getting closer to the end than they are to the beginning – little notions of where they are in life will be expressed.

But, in my case, I've noticed that not only is Time compressing and moving faster and faster in my life, but even the rate of acceleration is increasing just a tiny bit every day.

And, this is what we know is happening everywhere in the universe since the launch of the Hubble Telescope. We now know that the universe is expanding outwards in all directions AND, not only is it expanding in all directions but the rate of expansion is accelerating every minute and every second. This new information goes against all of our previously known laws of Gravity which should be pulling everything in the universe closer and closer which should be slowing the expansion of the universe.

This new knowledge about the universe has also brought about the discovery of 'Dark Matter' and 'Dark Energy' which appears to be responsible for this unexpected expansion in all directions as we have discussed.

Today, I'm wondering if there is possibly a correlation between how Time is accelerating in the rate of expansion of the universe and the rate at which our own time clocks expand as we age.

When I view this phenomenon from the perspective of what we've learned in this book, I am certain that there is such a correlation because – As the universe expands so does Consciousness. And, as we get closer and closer to the end of our time on this planet, not only does our knowledge and Consciousness expand, but the rate of expansion accelerates at the exact same rate as the universe expands – only on a smaller scale.

This can be expressed by a simple chart:.

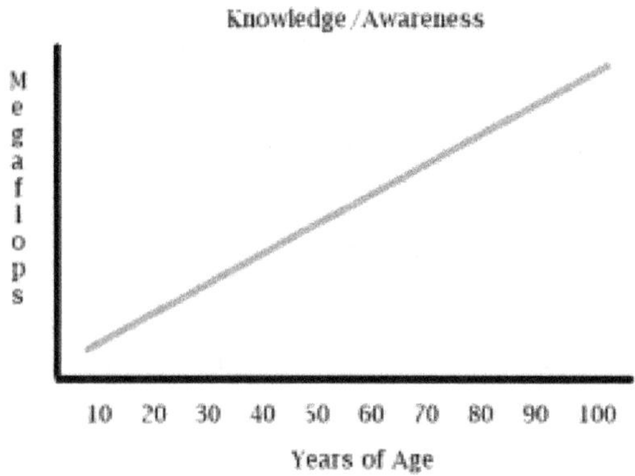

I believe that every life follows this format, though some lives will finish with only a slight increase in Consciousness, Awareness, Knowledge because they have decided not to try too hard. And so we've all known that type of personality where they just don't care about anything outside of themselves and all of their life's events are centered around selfish and meaningless acquisition of goods, wealth, acclaim, money.

But, even these poor lost souls will have attained some modest measure of an increase in their Consciousness about life, even though they were never even trying. A gain in awareness cannot be avoided completely since survival of even the lowest depends on it. Growth, even in the heart and soul of the lowliest of beasts must occur at some rate.

But, the most important part of this new discovery of ours is that there has to be a reason why we gain so much knowledge, awareness, Consciousness during our lives and that the total accumulated knowledge or Consciousness peaks at the moment of our leaving this physical body that we are born into. It's the same reasoning where we have placed the responsibility for the expansion of the universe to some unknown force that we call 'Dark Matter' or 'Dark Energy'. There has to be a force that is responsible for this great expansion and the increased acceleration of that expansion. Our Science requires that something must exist to perform this amazing phenomenon, the largest in our knowledge base of all things.

And so it must be the power of Consciousness itself that forces this rate of expansion upon us so that when we enter the next level of existence that our energy will convert into, we will enter it with a much higher and better vantage point. We know that energy is never destroyed, only transformed, and so it must go with the energy we have attained via our gray matter inside our skulls. No matter how bright our light bulb or how dim it may be, there is inside all of us about 150 millivolts of electricity generated inside our brain that operates solely to keep us alive every minute of every day and night. When our bodies finally fail us for the last time, the energy that kept all of our machinery operating has to go somewhere. (See 'Law of the Conservation of Energy'.)

Precisely where our energy ends up is not known even to me and this may be the greatest and longest lived mystery that we will ever know. So, I cannot at this time tell you where your energy ends up. But, I can tell you that you will arrive somewhere with a new purpose in life and a fresh start somewhere in this universe, or possibly in another one very far away.

Happily, I can also tell you that, based upon the information that you have gained here, something that we have known for centuries as 'God' does exist and does care about each and every one of us. This 'Caring' about us is not the same as the 'Caring' that we experience in this physical form, but there is a notation of our existence, a higher knowledge of our existence and where it fits in the overall picture, because we now know from the Second Law of Consciousness that everything is 'Connected', that includes all of my energy and all of yours, no matter where it is maintained or how it is expressed.

We also know from the First Law of Consciousness that there has to be a Creator who or which creates all things, events, purposes, etc. And we know from the Third Law of Consciousness that it consists in all things, even our own observations, activities, memories, life events and purposes. All of it, every last bit of it is stored as energy somewhere and so it is never destroyed, excect perhaps by the creator of it all who also maintains total control over all of the swirling energies of the universe. I can't prove that of course, but you should consider this a pretty good guess based on what we now know.

Therefore, my parting words to you as I have my final thoughts about all of this information is that you must make every waking moment of your life count, but it must count in the most prolific and productive way for something you truly believe in. Never give up hope that you can and must make a difference because this is our highest and best use of the time we are given and we are given it for that very reason.

You may have heard these words before, but I guarantee that you have never heard them in relation to all of the above where I believe I have proven the value of these words, the efficacy of these words and raised the importance of them to new levels.

For that I am most grateful.

YOUR FINAL SURVEY:

On a Scale of 1 - 5 with 1 Being the Lowest
and 5 Being the Highest . . .

How SATISFIED were you with your time on the Earth?

Very Dissatisfied	Somewhat Dissatisfied	Neutral	Somewhat Satisfied	Very Satisfied
O	O	O	O	O

I'll let you guess what my choice would be.

About The Author

Michael Mathiesen

Someday, I would like to be known as the discover or the 4 States of Consciousness as well as the Three Laws of Consciousness

---

I LIKE TO DESCRIBE myself as a life-long Science and Science Fiction student with a specialty in the area of Consciousness. With a college degree in Psychology, I was underwhelmed as an undergraduate working in an area with a total lack of any real research into the question of how Consciousness makes us think about things the way we do? And where Consciousness might reside in our brains? Or How it's constructed.

The personal computer revolution would start to yield some of this research though not from Academia, but out of Silicon Valley. As computers got smarter and smarter, they tended to force us to ask these questions until one day an artificial intelligence may emerge from a lab somewhere that can outthink even the smartest living individuals on the planet.

So, I wrote a book about that called - Metamorphosis: HashTag AI Cocoon

Being in the computer business and an early adopter of the Internet, I had plenty of time to think about this question. I studied

Genetics, Biology, Evolution, Computer Science, Economics, Quantum Physics, Astronomy, Crypto-Currencies and through it all I uncovered a pattern, a train of thought, a subtle yet living filament of Truth that had yet to be appreciated.

You can follow my lifelong line of inquiry from the many books I have authored along the way. The first major epiphany of mine came from the announcement of the discovery of the God Particle or Higgs boson from research done at the Large Hadron Collider near Geneva, Switzerland. This single energy is now considered to be the cornerstone of all modern understanding of the forces in the universe. Until July 4th, 2012, when they made this momentous announcement, we had no idea about the real nature of the Big Bang.

In The God Particle Bible – I lay out my theory that the God Particles were sent out FIRST like an advance attacking force from the Big Bang because since these invisible eager beavers give every other particle their mass, they would have to be first ones out of the gate in order to create the first electrons, protons, neutrons, quarks, etc that eventually congealed into the stars, galaxies, planets and eventually even us.

And so I say - "It's no accident that these particles were nick-named - 'God Particles.'"

From here, I started to think about all of the other energies from the God Particles on up.

The Electron, for example, somehow knows where to orbit the nucleus of every atom it joins in order to create. There are only a handful of distances away from the nucleus where they have permission to orbit. And, every single atom in the universe, as far as we know, and of which there is no number large enough to count them - contains these same exact places where the electron energy can live and no others.

But these places have long been known. More recently, however, it has been shown that electrons can also pair up, or become 'Entangled' with another electron. From this point forward, each electron in the pair knows about and responds to the existence of the other and so the question I then asked myself is a simple one: 'How do they know?' They're far too small to have brains or any other kind of neural network, as far as we know.

From these questions came the book - 'The 4 States of Consciousness' wherein I demonstrate how Consciousness begins at the subatomic level and is passed up through three more levels of Consciousness and into our own existence.

In a more recent book, 'The Science of Physics - Proof That God Exists', I go further and create a new number and even a simple formula based on my new number - that demonstrates how and even why the universe is constructed in this extremely intricate and intimate manner.

"I want to be remembered as the fella who discovered the 4 States of Consciousness and the Three Laws of Consciousness. That's all!"

If you're interested in learning more about how this story came to be, you will find clues in some of my earlier books such as:

The Science of Physics

The One of Hearts

The God Particle Bible

The 4 States of Consciousness

The Origin of Creation

The Blockchain Government

Metamorphosis: HashTag AI Cocoon

Extinction Live

Anti-Matter

Brain Drain

Maxtricity

Can Electrons Learn?

Zentanglements - The Three Laws of Consciousness for Smarties and many others -

All of which can be found at wherever you found this book or in retail book stores – just about anywhere.